RECOVERY, RECYCLE AND REUSE OF INDUSTRIAL WASTES

By Kenneth E. Noll, Ph.D., Charles N. Haas, Ph.D., Carol Schmidt, Prasad Kodukula
Pritzker Department of Environmental Engineering
Illinois Institute of Technology
Chicago, Illinois

INDUSTRIAL WASTE MANAGEMENT SERIES
James W. Patterson, Executive Editor

LEWIS PUBLISHERS, INC.
121 SOUTH MAIN STREET □ P.O. DRAWER 519 □ CHELSEA, MICHIGAN 48118

Library of Congress Cataloguing in Publication Data
Main entry under title:

Recovery, Recycle, and Reuse of Industrial Waste.

   (Industrial Waste Management Series   )
   Bibliography: p.
   Includes index.
   1. Recycling (Waste, etc.) 2. Factory and
trade waste. I. Noll, Kenneth E. II. Series.
TD794.5.R415   1985   363.7′28   84-23383
ISBN 0-87371-002-9

2nd Printing 1985

LEWIS PUBLISHERS, INC.
121 South Main Street, Chelsea, Michigan 48118

PRINTED IN THE UNITED STATES OF AMERICA

This volume, <u>Recovery Recycle and Reuse of Industrial Waste</u>, is directed to an audience including industrial and environmental engineers, and managers. It is designed to explain the underlying concepts, advantages and, in certain instances, disadvantages of recovery, recycle and reuse; to provide examples of existing applications of recovery technologies; and to identify and evaluate applications of a broad spectrum of existing and potential recovery technologies. Of particular value for the technologist is the organizational format of the technologies discussed, as well as the information on potential applications and, where they exist, current technical limitations on those applications. Managers will benefit from the non-technical overviews provided, plus the discussions of economic implications of recovery, recycle and reuse.

Review of this work by the USEPA does not necessarily indicate approval of its content. Special appreciation is expressed to Mr. William A. Cawley, Deputy Director, USEPA Industrial Environmental Research Laboratory, and Chairman of the IWERC Policy Board.

James W. Patterson, Ph.D.
Series Executive Editor and
Director of
The Industrial Waste Elimination
Research Center

SERIES PREFACE

The philosophy of industrial waste management
has changed rapidly in recent years, due to regulatory and economic incentives resulting from increasing restrictions on air, water and hazardous
wastes pollution, and from escalating costs of
pollution abatement. Traditional methods of industrial pollution control often ignore alternative
management strategies with significant economic and
environmental benefits. Specifically, opportunities for source reduction, by-product recovery, and
recycle and recovery for reuse have historically
received scant attention. Today, industry is increasingly considering such alternative approaches,
and finding many opportunities for their cost effective implementation.

The Industrial Waste Elimination Research
Center (IWERC) is a consortium of Illinois
Institute of Technology and the University of Notre
Dame, established under the auspices of the U.S.
Environmental Protection Agency (USEPA), Office of
Exploratory Research. It operates under the super-
vision of a Policy Board, and with advice from its
Scientific Advisory Committee and Industry Advisory
Council. IWERC has the mission to identify manage-
ment strategies, and to perform technology evalua-
tion, research, and development on innovative alternative technologies for industrial waste management. As part of its program, the Center commis-
sioned comprehensive studies in three areas:

1.   Recovery, recycle and reuse of
      industrial pollutants;

2.   Management of industrial pollutants by
      anaerobic processes; and

3.   Process modifications for pollution
      source reduction in chemical processing
      industries.

These studies have yielded state-of-the-art information on opportunities for applications of alternative strategies for industrial waste management.
In view of the current widespread and growing interest in such alternative approaches, the results
of these studies are being made available, as part
of this Industrial Waste Management Series.

James W. Patterson, Ph.D.

CONTENTS

INTRODUCTION

Increased emphasis has been placed on studies
of the chemistry, biological effects, treatment,
fate, and control of industrial by-product pollutants. Discovery of the presence of such materials
at high concentrations coupled with the recognition
of their environmental impacts and potential health
hazards has led to major legislative efforts which
would limit their release into the environment. In
addition, there is an increased interest to ensure
the continued viability of domestic minerals (which
constitute most of the raw materials for various
industrial operations), minerals economy, and the
maintenance of an adequate mineral base. The environmental regulations prohibiting the discharge
of toxic pollutants from industrial activities,
coupled with the need for conservation of raw
materials has led to consideration of the recycle,
recovery, and reuse of waste products. The
recycle, recovery, and reuse alternative is doubly
advantageous since it conserves a materials supply
which is beginning to be recognized as finite,
while reducing the quantity of hazardous pollutants
discharged into the environment.

The choice between recycle, recovery and reuse
of valuable materials from waste and disposal of
waste seems to depend mainly on two factors; economics and technology. \Economics is probably the
most important factor that limits the recycle,
recovery, and reuse of industrial by-products.\ The
high cost of recovering low-value materials and the
consequent relative unprofitability seem to prevent
many industries from adoption of recycle or
recovery techniques for waste by-products. The
high costs of recycle or recovery techniques,
however, can probably be reduced by improving the

available technology. Any such attempts would
require the identification of technological limitations associated with the recycle and recovery
techniques. Therefore, it is necessary to collect,
review, and systematically organize and evaluate
information pertaining to the state-of-the-art for
recycle, reuse and recovery of by-product pollutants; the efficiency, energy, and resources
associated with these processes; and the future
needs and demands for reduction, elimination, or
reuse of the unwanted by-products. The execution
of these processes results in a collection of information which defines the limiting technology and
the energy and economic constraints associated with
various techniques and processes, as well as their
potential for future development and expansion as
valuable waste elimination methods. Such information is of use to focus future developments in the
area of recycle, recovery, and reuse technology
aimed at the elimination of industrial by-product
waste.

OBJECTIVES

     The major goal of this book is to produce a
document useful in planning future efforts aimed at
the elimination of industrial wastes through the
application of recycle, recovery, and reuse technology. This objective was accomplished by collecting, reviewing, and evaluating information pertaining to the state-of-the-art for recycle, recovery,
and reuse of by-product pollutants using different
industrial processes. From this information, conclusions were made regarding the technological
limitations associated with recycle, recovery, and
reuse of industrial pollutants.

     This study is not meant to provide detailed
technical information on the treatment process
itself nor elaborate discussions on the various
applications of the processes. It does however,
provide an overview of the applications of the
various processes to the recovery of contaminants
which may subsequently be recycled or reused.

ORGANIZATION OF THE BOOK

     In this book, an attempt is made to identify
technological limitations associated with unit
processes currently or potentially employed for
separation of organic and inorganic pollutants from

various industrial waste streams. This goal was
accomplished by collecting, reviewing, evaluating,
and summarizing information pertaining to the
state-of-the-art for the recycle, recovery and
reuse of pollutants in industrial waste effluents.
The pollutants considered in this study are
basically organic and inorganic by-products from
wastewater effluents, solid residue and gaseous
emission from industrial operations.

The information collected during the course of
this project is summarized and presented under two
sections. The first section of the book consists
of chapters on the methodologies currently available for recovery of industrial and hazardous
waste, and developing technology for recycle,
reuse, and recovery. The second section of the
book contains chapters concerned with 5 technical
categories used for recovery; namely, sorption,
molecular separation, phase transition, chemical
modification, and physical dispersion and separation. This categorization is based on the type of
transformation a pollutant would undergo during
recovery in a given unit process or operation.

In each chapter in the second section, individual unit processes are presented with a brief
discussion of the process itself, its applications,
and the technological limitations associated with
the process. If a process is used for removal of
both organic and inorganic pollutants (for example,
ion exchange), its application and the technological limitations are discussed in the same chapter
but under different sections for each category of
pollutant. A similar approach is adopted in the
case of processes (for example, adsorption) which
are based on principles common to the removal of
pollutants from both gaseous and wastewater
streams.

At the outset, each chapter provides a process
description. This would include the underlying
principles of the process, operating characteristics, process configurations, and any unique features of the process.

Immediately following the process description
section, each chapter discusses the limiting technology associated with the concerned treatment
process. Conclusions on technological limitations
of the process for by-product recovery and reuse

are made based upon the literature information
gathered on the process. Attempts were made to
make such conclusions as specific as possible. One
of the major problems encountered was the literature pertaining to the processes and techniques
employed in recovering industrial by-products which
were of a general and somewhat non-technical nature. It was consequently sometimes difficult to
make specific conclusions on limiting technology
based on available technical and performance data.
As can be expected, due to proprietary value, technical information on industrial processes is not as
extensively available as it is in the case of
common waste treatment processes.

The next section of each chapter deals with
application of each process for pollutant removal
and recovery from industrial waste streams. The
types of industries using the process for pollutants removal are listed in this section. In a
number of cases, recovery of by-products is not
practiced. It seems that very little technical
data pertaining to the recovery of by-products in
real-life industrial situation are available. In
most cases, even the process performance data are
scarce due to the proprietory value of the
material. It is also possible that information on
recovery of industrial by-products is scant due to
the fact that this area has received attention only
in recent years. The last section of each chapter
gives a list of the references used in preparing
the chapter.

WASTE CLASSIFICATION FOR RECYCLE, RECOVERY AND REUSE

The quantity and quality of industrial wastes
generated by various industries is difficult to
identify; however, Jennings (1982) has provided an
initial evaluation by conducting a national industrial residual flow study. Table 1.1 presents the
rank order of industries producing residues. Industrial residuals were defined as those residues
that are routed either to unique treatment technologies or to chemical waste disposal facilities and
considered "hazardous" under the Resource Recovery
Act and The Toxic Substances Control Act. Current
data indicate that these constitute about 50 percent by weight of the total industrial discharge.
The remaining fraction is composed of such items as
industrial trash, foundry sand, wood waste and

TABLE 1.1

TOTAL QUANTITY OF INDUSTRIAL WASTE ORDERED BY
MANUFACTURING INDUSTRIES THAT PRODUCE HAZARDOUS
RESIDUAL

INDUSTRIES	PERCENT OF TOTAL QUANTITY*	
CHEMICALS	37.6	
PRIMARY METALS	29.1	74.4
FABRICATED METALS	7.7	
MACHINERY	6.5	
PAPER	4.6	
TRANSPORTATION	4.0	
FOOD	2.7	22.9
PETROLEUM	2.4	
STONE	2.0	
ELECTRICAL	0.7	
RUBBER	0.7	
LEATHER	0.5	
LUMBER	0.4	
INSTRUMENTS	0.3	
MISC. MANUFACTURING	0.2	
FURNITURE	0.2	2.6
TEXTILES	0.2	
PRINTING	0.1	
TOBACCO	<0.1	
APPAREL	<0.1	

*TOTAL = 100% of 27.8 million tons of wastes
reported from 21 states.

slags.

Table 1.1 shows that 75 percent of the residual volume is from the chemical, primary metals,
and fabricated metals industries. Jennings was
also able to estimate the percentage of residual
material as to solids, liquid, or sludges, as shown
in Table 1.2. Under each category, the residuals
were identified by physical and chemical properties
(Table 1.3). These tables show that liquids and

TABLE 1.2

PERCENTAGE OF RESIDUAL TYPE IN
EACH LEVEL I WASTE CLASSIFICATION

CLASSIFICATION	PERCENT OF TOTAL* IN CLASS
SOLIDS	13.9
LIQUIDS	54.4
SLUDGES	23.8
UNIDENTIFIED	7.9

*Total = 100% of 15.4 million tons of
       hazardous wastes reported from
       30 states.

sludges are a large percent of the total residue.
This remains true even when the unidentified category is lumped with the solids. The miscellaneous
special solids category contains wastes such as
pesticide solids and containers, explosives, patho-
genic wastes, DOT "poisons" and similar residues.
The metal solutions and metal sludge categories
contain predominantly heavy metal residuals. The
liquid categories contain both dilute and concentrated solutions and (where appropriate) nonaqueous liquids. Metals solutions and metal
sludges accounted for nearly 15 percent of the
total.

METHODOLOGY

For the purpose of clarity, it is necessary to

TABLE 1.3

PERCENTAGE OF RESIDUAL TYPE BY CHEMICAL
CHARACTERISTICS

SOLIDS	PERCENT OF TOTAL[1] IN CLASS
ORGANIC SOLIDS	11.8
INORGANIC SOLIDS	28.0
MISCELLANEOUS SPECIAL WASTES	60.2

LIQUIDS	PERCENT OF TOTAL[2] IN CLASS
HALOGENATED ORGANICS	1.0
NON-HALOGENATED ORGANICS	8.6
ACIDS	34.8
CAUSTICS	18.8
METAL SOLUTIONS	11.4
OILS & OILY WASTES	7.6
MISCELLANEOUS LIQUIDS	17.8

SLUDGES	PERCENT OF TOTAL[3] IN CLASS
ORGANIC SLUDGES	41.8
INORGANIC SLUDGES	33.5
METAL SLUDGES	24.7

1. Total = 100% of 2.1 million tons of hazardous
         wastes reported from 21 states.
2. Total = 100% of 8.2 million tons of hazardous
         wastes reported from 23 states.
3. Total = 100% of 3.6 million tons of hazardous
         wastes reported from 19 states.

define the terms recovery, reuse, and recycle. Recovery is defined as the extraction of any pollutant from wastes. The terms reuse and recycle describe the manner in which the recovered materials are put to use. Reuse of the material is its utilization for any purpose, whether it is the same or different from its previous use. Recycle will be defined as a specific type of reuse in which the recovered material is reused for the same purpose as that for which it had been used previously.

The effort for this book was divided into two different phases and a summary of these phases is in order.

Phase I.   Literature Collection

This phase consisted of a thorough collection of literature dealing with information on technology which has or may have application in the areas of recycle, recovery and reuse of by-product wastes, as well as information directly related to the industrial by-product materials themselves.

The literature search was approached with the following basic questions in mind: (1) What wastes are presently produced as industrial by-products? (2) What methods can be applied to these wastes to bring about their recycle, recovery and reuse?

Many sources are available which contain information pertinent to the project, and the tools to facilitate its collection. Sources include journals, reports, books, and the proceedings of conferences and symposia. In order to amass the largest collection of relevant information in the short period of time available, it was necessary to make use of on-line computer literature searches, published literature searches, collections of abstracts, and the literature reviews which appear periodically in certain journals.

On-line computer literature searches (for example, The EPA Computerized Literature Search System) were used to access data bases and provide bibliographic information in response to specific questions. Published literature searches such as those produced by The National Technical Information Service (NTIS) were valuable during the course of the project. Over 1,200 abstracts concerning recycle, recovery and reuse were collected from the

Engineering Index.

Phase II. Literature Categorization

During this phase, the abstracts collected from Phase I were reviewed and articles relevant to the project objectives were collected. Information collected from computer searches and published bibliographies were carefully reviewed and pertinent articles obtained.

A review of collected literature indicated that the categorization of literature can be based upon one of the three following approaches: a) industry-by-industry, b) process-by-process, c) material-by-material.

An industry-by-industry approach would have required a classification of industries and their respective wastes and an evaluation of the several possible methods by which these wastes could be treated. It seemed that several industries using a given treatment process had similar technological limitations in terms of recovery of their by-product pollutants. Thus, the same conclusions on limiting technology would have to be repeated for a number of industries for a given treatment process.

If the literature categorization had been based on materials, the waste by-products would require initial identification followed by studies of various processes suited for their recovery and reuse. A careful evaluation of the collected literature indicated that the limitations of technology are more dependent upon the treatment process itself than on the type of waste or type of pollutant treated. It was, therefore, decided that for the purpose of this project, the categorization of literature would be based on treatment processes.

A schematic of literature categorization used in this project is shown in Figure 1.1. The literature was broadly divided into three major areas; air, water, and solids, each of which deals with two major classes of by-products, namely, organics and inorganics.

A total of 18 processes which are believed to have the most widespread application in terms of removal and recovery of industrial by-products were

FIGURE 1.   LITERATURE CATEGORIZATION

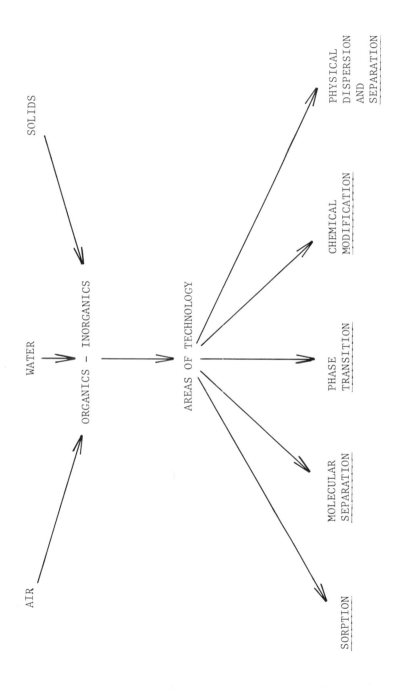

selected for study in this project. A list of the
processes is presented in Table 1.4. These processes are divided into 5 groups, depending upon
the type of transformation the by-product would
undergo during the waste treatment in a given unit
process or operation. The five groups of processes
are: sorption, molecular separation, phase transition, chemical modification, and physical dispersion and separation.

After the identification of the unit processes
listed in Table 1.4, a research specialist was
assigned to a unit process. Each treatment process
was evaluated according to the following outline:

I.   The Process
II. Limiting Technology
III. Recycle, Recovery, and Reuse
     Applications
IV. References

It should be pointed out that the references
given at the end of each chapter are listed in
alphabetical order. Attempts were made to obtain
as many recent references as possible. The references are classified into four categories:

1)   Articles referenced: The articles
     listed under this category are actually
     referenced in the text.

2)   Articles read: In this case the
     original articles were read.

3)   Abstracts read: Under this category of
     references only the abstracts of the
     articles were read.

4)   Titles read: Only the titles of the
     articles were read for the references
     under this category. No attempts to
     obtain the abstracts or the original
     articles were made.

The references listed under the last three
categories are not always quoted in the actual text
portion of the chapters. These are, however, included in the references section of each chapter
because it was felt that they are related to the
information presented in the text and the reader
may refer to these, if interested.

Table 1.4   Treatment Processes Selected

SORPTION	MOLECULAR SEPARATION	PHASE TRANSITION	CHEMICAL MODIFICATION	PHYSICAL DISPERSION AND SEPARATION
ABSORPTION	REVERSE OSMOSIS	DISTILLATION	CHEMICAL REACTIONS	DEWATERING
ADSORPTION	ION EXCHANGE	EVAPORATION	THERMAL REACTIONS	FILTRATION
	ULTRAFILTRATION	CONDENSATION	CEMENTATION	FLOTATION
		REFRIGERATION		DROPLET SCRUBBING
				EMULSION
				LIQ-LIQ EXTRACTION

SUMMARY AND CONCLUSION

In this study an attempt was made to collect, review, and systematically organize and evaluate information pertaining to the state-of-the-art for recycle, reuse, and recovery of industrial pollutants through the use of various waste treatment processes, and to suggest technological limitations associated with these processes.

At the present time, economics seem to be a major factor in determining if recylce, recovery and reuse of industrial pollutants are appropriate. However, current economic costs do not include the "hidden" costs to society of environmental pollution and depletion of natural resources. If the waste treatment technology is improved to a level where economics does not become the limiting problem, recycle, recovery and reuse of industrial pollutants could become widespread.

Generally, most of the processes discussed in this report are well-developed and proven to be successful for effective removal of pollutants they are designed to treat. However, a majority of processes are not either economically or technically effective for recovery of by-product pollutants. For example, precipitation is an effective process for removal but not for recovery of heavy metals. Many industrial wastes receive precipitation treatment but, in most cases, the resulting sludges are disposed of without recovery of metals because of economic constraints.

In the case of processes such as evaporation, there seem to be no insurmountable technical problems, however, economics plus operational difficulties such as crystal formation and scaling seem to limit the recovery application of the process in many industrial operations.

Some of the processes (e.g., cementation) studied in this project are at an infant stage while some others are very well established (e.g., activated carbon). In the case of cementation, the process is well developed and commonly used in hydrometallurgical operations. However, its application to industry to date has been limited, and there is a large scope for research in this area. Even though major advances have been made in the area of activated carbon technology, fundamental

research on adsorption/desorption mechanisms is
warranted, which would lead to improved removal and
recovery of organic and inorganic compounds.

The unit processes considered under the category of sorption are highly developed for removal
of organics and inorganics from both gaseous and
wastewater effluents. The process of absorption
for removing and recovering gaseous pollutants
seems to be most efficient when the pollutants are
quite soluble in the absorbent and when the absorbent is relatively non-volatile, non-corrosive, and
has low viscosity. The rates of mass transfer
between the gas and the absorbent are primarily
determined by the amount of surface area available
for contact. Contact between the gas and liquid
solvent are most commonly achieved in plate and
packed towers. Absorption of pollutants (such as
phenolic compounds, hydrocarbons, H_2S, SO_2, etc.)
followed by desorption constitutes a cyclic operation which allows reuse of the sorbent and acts as
a device for separation and concentration of the
selected pollutant.

The process of adsorption in the area of air
pollution is used for removal of odors, hydrocarbons, Hg, SO_2, and No. The principal adsorbents
are activated carbon, silica gel, activated
alumina, and synthetic zeolites. In the recovery
of organic gaseous pollutants, activated carbon has
been by far the most effective adsorbent used. The
recovery of organic compunds after breakthrough has
been reached is generally achieved by stripping.
The conventional methods of regeneration of air
pollutants involve heated air, heated inert gas or
heated steam, whereas thermal reactivation, steam
regeneration, and acid or alkaline regeneration are
used for recovery of organics or inorganics from
adsorbates used in wastewater treatment.

The technological limitations associated with
sorption process are summarized in Table 1.5.

Of the three processes considered under molecular separation, ion-exchange is a well established process and has been used in industrial operations for a long time whereas, membrane processes
such as reverse osmosis and ultrafiltration have
received attention only in recent years.

TABLE 1.5   LIMITING TECHNOLOGY FOR SORPTION
            PROCESSES

ABSORPTION:

'   SOLUBILITY OF GAS IN THE LIQUID

ADSORPTION:

                UNDERSTANDING OF FUNDAMENTAL MECHANISMS
                OF REMOVAL

                REGENERATION PROCESSES

                LACK OF COMPREHENSIVE PREDICTIVE MODELS

During the process of ion exchange, undesirable ions from a waste stream are transferred on to the ion exchange material, such as synthetic resin, which may be regenerated. Through regeneration, the ions may be recovered in a concentrated stream, and the ion exchange material is made available for further use. Recovery of chromic acid and valuable metals such as chromium, nickel, silver and gold from plating industry effluents is possible. Resins of the macromolecular type have become favored in recent years because of their relatively higher resistance to fouling.

Membrane processes are operated by hydraulic pressure and remove solution components largely on the basis of molecular size and shape involving neither a phase change nor interphase mass transfer. These processes are used to removal hazardous materials contained in both air and liquid waste streams.

Reverse osmosis is generally applied for removal of low molecular weight solutes, such as salts, sugars and simple acids from their solvent, whereas ultrafiltration is used for separation of higher molecular weight solutes such as proteins, starch, natural gums and colloidal materials such as clays, pigments and microorganisms. The membranes are made from other synthetic or natural polymeric materials.

There appears to be a submstantial need for improved membrane technology. Table 1.6 presents the major technological limitations associated with processes studied under the category of molecular separation.

The processes considered under the category of phase transition are evaporation, distillation, condensation and refrigeration. The first two processes are used for the removal and recovery of wastewater pollutants whereas the latter are for gaseous pollutants. In some metal and plastics finishing industries and, particularly the electro-plating industry, closed-loop recycling of wastes is achieved by using evaporation as a recovery process. Evaporation is a very costly process; however, it becomes cost-effective as the concen-trations of pollutants in the wastes increase and the flow rate becomes low. Multiple-effect evapo-rators are more common in the electroplating indus-try because of their relatively lower operating costs.

TABLE 1.6   LIMITING TECHNOLOGY FOR MOLECULAR
            SEPARATION PROCESSES

REVERSE OSMOSIS, ULTRAFILTRATION AND ION EXCHANGE TECHNOLOGY IS LIMITED BY THE BEHAVIOR OF THE SEPARATED MATERIAL AND ITS EFFECT ON PROCESS PERFORMANCE. FURTHER STUDY OF SEPARATION THEORY AND BETTER UNDERSTANDING OF THE CAUSES OF PROCESS FAILURE ARE NEEDED.

Distillation is a process in which the vapori-zation of a liquid mixture yields a vapor phase containing more than one component. As a unit operation, it has been used successfully either singly or in combination with such operations as direct condensation, adsorption, and absorption for the recovery of organic solvents. Distillation has many applications for compound recovery from indus-trial wastes. The regeneration of activated carbon may result in a liquid which is distillable for recovery of the organic component. Other applica-tions include recovery of methylene chloride from polyurethane waste, the recovery of organics from plating wastes and, the recovery of waste solvents for reuse in cleaning industrial equipment.

The condensation process is employed by chemical process industries to recover solvents and products which can be recycled to manufacturing processes, and is also used to recover volatile hydrocarbons from fuel-storage operations. It employs either contact or non-contact methods for cooling a vapor to the point where the partial pressure of the condensible component equals its vapor pressure. Less commonly, the temperature of the system is held constant and the system pressure is increased until the component's partial pressure equals its vapor pressure.

The production of cooling (or heat withdrawal) may be accomplished by the solution melting, or evaporation of a substance, or by the extension of a gas. The term refrigeration refers particularly to cooling below atmospheric temperature. Refrigeration is one of several competing methods for recovering emissions from bulk liquid transfer and storage operations, and has been promoted for vapor recovery at gasoline loading racks.

Table 1.7 presents the technological limitations of processes considered under the category of phase transition.

TABLE 1.7   LIMITING TECHNOLOGY FOR PHASE TRANSITION
            PROCESSES

EVAPORATION:

> CRYSTAL FORMATION, SALTING, SCALING, CORROSION ENTRAINMENT, AND FOAMING

DISTILLATION:

> ENTRAINMENT EFFECTS ON ATTAINABLE PURITY

CONDENSATION:

> LOW REMOVAL EFFICIENCY AT LOW LEVELS OF CONDENSIBLE VAPORS

> FOULING OF HEAT EXCHANGE SURFACES BY PARTICULATES

Chemical precipitation and reduction are commonly used to remove and recover metals from industrial waste effluents. Whereas the former process is used in a number of industrial operations to remove various heavy metals, chemical reduction is mostly used to reduce hexavalent chromium to its trivalent form in the plating and tanning industries. The reduction process in this case is followed by precipitation of trivalent chromium with either lime or sodium hydroxide.

The cementation process, which is widely used in hdyrometallurgical operations, is not well developed for the removal and recovery of metals from industrial waste streams. In this process, the ionized metal in solution is converted to its elemental stage by spontaneous electrochemical reaction through oxidation of another elemental metal which is also kept in solution. The process performance can be predicted in terms of electrode potentials. This process is presently in its infant stage and there seems to be a large scope for research and full-scale application in this area.

Catalytic hydrogenation is a useful method for achieving controlled transformation of organic compounds. Using this technique, it is possible to saturate organic compounds with hydrogen. Catalytic hydrogenation has been applied in the production of substitute natural gas having a high concentration of methane and ethane. It has also been used to recover hydrogen gas, and to convert the sulfur present in tail gas to hydrogen sulfide which can undergo conversion to elemental sulfur.

A summary of technological limitations associated with the processes involving chemical modifications of pollutants treated is presented in Table 1.8.

The last category of unit processes considered in this study is physical dispersion and separation. All the unit processes included in this category (except for liquid-liquid extraction) are used for removing suspended matter from wastewater or gaseous streams.

Filtration is a physical process in which solids suspended in a gaseous or a liquid stream are separated by passage through a previous medium that separates and retains either on its surface or

TABLE 1.8   LIMITING TECHNOLOGY FOR PROCESSES
INVOLVING CHEMICAL MODIFICATIONS

CEMENTATION:

> NOT EFFECTIVE FOR LARGE FLOW DUE TO
> LONGER CONTACT TIMES REQUIRED

> EXCESS IRON CONSUMPTION RESULTING IN
> ESCESS IRON SLUDGE REQUIRING DISPOSAL

> THERMODYNAMIC LIMITATIONS FOR ACHIEVING
> THE DESIRED LOW LEVELS

PRECIPITATION:

> PRESENCE OF COMPLEXING AGENTS

> ACHIEVING OPTIMAL REMOVALS FOR MORE THAN
> ONE METAL AT ONE pH

REDUCTION:

> HARD TO ACHIEVE INTIMATE CONTACT BETWEEN
> THE REDUCTANT AND THE POLLUTANT,
> ESPECIALLY IN CONCENTRATED WASTES

> INTRODUCTION OF NEW METAL IONS WHICH
> NEED FURTHER TREATMENT FOR REMOVAL

PYROLYSIS:

> LACK OF UNDERSTANDING OF EFFECTS OF
> PROCESS VARIABLES ON PROCESS PERFORMANCE

CATALYTIC HYDROGENATION:

> RATE OF CHEMICAL REACTION

> RATE OF HYDROGEN TRANSPORT

---

within itself, the solids present in the suspension. In all filtration processes, a pressure differential is induced across the medium to force the gaseous or liquid stream to flow through it.

In the case of filtration of liquids, either surface filters in which the solids are deposited on the upstream side of the medium, or deep bed filters, which deposit solids within the medium, are used. Surface filters are normally used for suspensions with more than one percent solids whereas dilute suspensions are treated by deep-bed filters. A wide variety of filtration devices are commerically available.

The filter media used for liquid filtration may be a filter cloth, filter screen, a layer of granular media such as sand, lake, coal, or porous ceramics or a barrier composed of a disposable material such as powdered diatomaceous earth or waste ash. Filter media used for filtration of gases may be porous paper, woven and felt fabric filter, or gravel or sand aggregate beds. Filtration has been used for removal of suspended matter from innumerable types of industrial wastewaters and gaseous streams. Multi-media filtration is commonly used for removal of the metal precipitates from wastewater after it has been subjected.

Flotation is a unit operation used to separate solid or liquid particles from liquid phase. In this process, fine air bubbles which are introduced into the liquid phase attach to the particulate matter and, consequently, the buoyant force of the combined particle and gas bubbles becomes large enough to cause the particle to rise to the surface. Three methods of introducing gas bubbles are available: dissolved air flotation, induced air flotation and vacuum flotation.

The technological limitations associated with processes under the category of physical dispersion and separation are presented in Table 1.9.

The process of liquid-liquid extraction involves separating the components of a liquid-liquid mixture by the addition of another liquid referred to as the solvent, which is immiscible (or only partially miscible) with the initial phase. The solvent is chosen so that one or more of the components of the original solution will transfer preferentially into the solvent phase leaving the others behind in the raffinate. The major applications of this process in wastewater treatment engineering are 1) recovery of phenol and related compounds from wastewaters and 2) removal of water soluble solvents such as alcohol from wastes containing mixed chlorinated hydrocarbon solvents.

TABLE 1.9   LIMITING TECHNOLOGY FOR PHYSICAL
             DISPERSION AND SEPARATION PROCESSES

FILTRATION:

>   LACK OF UNDERSTANDING OF PARTICLE/FILTER
>   INTERACTION
>
>   MAINTENANCE OF FILTER IN GOOD CONDITION

LIQ-LIQ SEPARATION:

>   SELECTION OF SOLVENTS AND CONTACTORS TO
>   PRODUCE DESIRED RESULTS

REFERENCES

Jennings, A.A., "Analysis of The National Inustrial Residual Flow Problem, IWERC Project 8001, 1982.

CHAPTER 2

INDUSTRIAL TREATMENT OF WASTE
FOR RECOVERY, RECYCLE OR REUSE

INTRODUCTION

Treatment of waste for recovery, recycle, or
reuse may be by chemical or physical techniques,
and may consist of a sequence of operations. A
list of separation methods for liquids and solids
is provided in Tables 2.1 and 2.2. The list has
been divided into chemical and physical methods and
is only an example of the possible methods that can
be combined to recover material. Recoverable materials may exist in a waste stream in many different
forms, for example, as metals or metal oxides.
They may exist as acid solutions of metal salts.
Therefore, the selection of a recycle process depends on specific character of the waste and requires careful consideration. For the purpose of
this work, technologies were grouped by their simi-
larities and not by their ability to accept similar
wastes.

INDUSTRIAL POLLUTION RECOVERY

EPA funded a study of alternatives to conventional pollution control that involves recycle,
recovery, and reuse [MRI, 1980]. Twenty-five case
reports obtained from the pollution contol literature for the 1974-1979 period are summarized. The
cases come from 11 industries, domestic as well as
foreign operations, and from small as well as
large companies and one public sanitary district.
A roughly equal number of cases deal with water and
air-related cases. Half the cases involve process
modification; the rest involve other types of optimization approaches. Most of the pollution control
by process change take place in the process-inten-
sive industries, such as chemicals, paper,

Table 2.1   TREATMENT METHODS FOR LIQUID EFFLUENT

Chemical Treatment Methods	Example
Absorption	Solvent recovery
Cementation	Copper recovery
Chlorination	Cyanide oxidation
Demulsification	Soluble oil recovery
Electrolytic processes	Metal recovery
Hydrolysis	Cellulose waste
Incineration	Waste oils
Ion exchange	Metal recovery
Neutralisation	Waste acid
Oxidation	Phenol removal
Precipitation	Metals
Reduction	Hexavalent chromium
Physical treatment methods	
Absorption	Removal of volatile organics
Crystallisation	Recovery of inorganic salts
Distillation	Solvent recovery
Evaporation	Sulphuric acid recovery
Filtration	Sewage sludge
Flotation	Dairy wastes
Foam fractionation	Metal separation
Phase separation	Oily wastes
Reverse osmosis	Desalination
Solvent extraction	Metal recovery
Stripping	Ammonia removal

Table 2.2 TREATMENT METHODS FOR SOLIDS WASTE

Chemical Treatment Methods	Examples/comments
Calcination	Gypsum
Chlorination	Tin removal
Cooking	Inedible offal
Froth flotation	Coal recovery
	Glass
Leaching	Gives an aqueous solution which may be treated as a liquid waste (Table 2.1)
Sintering	Colliery spoil
	Millscale
Physical treatment methods	
Centrifugation	Animal oil separation
	De-oiling swarf
Comminution	Mining wastes
	Cars
Drying	Filter cake
Granulation	Slag
Magnetic treatment	Iron removal from slag
Screening	Clinker

petroleum, metals, and food.

The common characteristic of the alternative approaches is change in the production process, resulting in reduction in elimination rather than the control of pollutants. Wherever pollutants can be eliminated--through changes in the production process, recycling or reuse of captured wastes, recirculation of streams, or some other strategy-- the need for, and the substantial costs of, terminal treatment systems are minimized or avoided.

Identifying characteristics of alternative approaches involve:

1. Recovery and reutilization of waste materials in the process itself, in the production operations that the system serves, or in an unrelated operation;

2. And/or energy conservation components based on overall design, utilization of process heat, energy generation from waste combustion, or combinations of these;

3. And/or use of substitute raw materials in the production process upstream of the "alternative system" in order to make the system more effective;

4. And/or modification of the production process for the same reasons.

Seven categories of innovation accommodate all the 25 cases reviewed. These are shown below:

New device
New procedure
New process
Modified process
Process chemical change/elimination
Fuel substitution
Other waste utilization

The largest number of cases involved process modification and changes in process chemicals, but process chemical change/elimination, in this sampling of cases, tends to be linked to other techniques. Fuel substitution, in each case, involved the use of plant or process wastes as a fuel; if

TABLE 2.3

Project	New/Modified Device	New Procedure	New Process	Process Modification	Change/Elimin. of Chemical	Fuel Substitution	Other Waste Utilization	Water Pollution Reduction	Air Pollution Reduction	Less Sludge	Less Solid Waste	Energy Conservation	Water Conservation	Raw Materials Conservation	Small Treatment System	Industry
Oil separator	X							X	X					X		Metals fabrication
Scrubber-absorber	X			X					X	X		X		X		Pulp & Paper
Catalytic oxidizer	X								X							Plastics
Fat separator	X	X		X				X						X		Food
Water conservation		X		X	X			X					X			Electronics
Fiber recovery				X				X						X		Pulp & Paper
Zero discharge			X	X	X			X				X		X		Construction Products
Steel sludge recovery			X	X				X		X		X		X		Iron & Steel
Pickle liquor recovery				X	X			X		X		X		X		Iron & Steel
Sulfur & energy recovery				X					X			X		X		Pulp & Paper
Water conservation				X				X				X	X			Pulp & Paper
Air pollution, energy control	X								X			X				Packaging
Fume incineration									X	X				X		Textiles
Acid recovery				X	X			X						X	X	Electronics
Odor control				X					X					X	X	Chemicals
Water conservation				X									X		X	Chemicals
Waste reduction				X		X				X				X		Chemicals
Water conservation								X	X				X			Petroleum
Solvent reclamation					X	X		X						X		Electronics
Acid elimination					X	X								X		Electronics
Sludge combustion										X	X	X				Waterworks
Metal recovery						X	X			X	X	X		X		Plastics
Energy recovery											X	X				Construction Production
Energy recovery											X	X				Pulp & Paper
Wastewater use							X						X			Pulp & Paper

this category is combined with other cases of waste
utilization, it becomes the second most common
category. In a large proportion of cases (23 of
25), benefits were reportedly realized in 2 or
more categoreis. In all but 1 case, the
costs of the innovation were lower than the
conventional solution by terminal treatment system.
An overview of the cases is presented in Table 2.3
showing the technique or techniques used. Six of
these cases resulted in recovery or recycle of
material and provide a good example of the diverse
nature of the technology.

1.   Fiber recovery. In this case a
      manufacturer of corrugating medium
      launched a campaign to capture and to
      reuse rejected fibers from four
      different operations. In each case,
      rejected fibers were captured,
      reprocessed, and reintroduced into the
      process at different stages. Resulting
      final products were tested to determine
      whether or not the addition of rejects
      adversely affected product quality. In
      the production of corrgating medium,
      waste fibers cause fewer problems than
      in fine paper or Kraft specialty
      production.

2.   Recovery of steel plant sludges. Large
      quantities of water are used in hot strip
      rolling mills in the still industry for
      removing fine scale from the metal in
      forming. Very fine oxides, together
      with hydraulic oils and lubricating
      greases released in rolling operations,
      enter the flume beneath rolling opera-
      tions with the water. The very fine
      particles of oxide are coatedwith oils
      and greases, and since small size means
      more surface area per unit of weight,
      the metals are very contaminated with
      oil. The oxides are too fine for direct
      recharging to furnaces and need to be
      agglomerated by sintering. The heat of
      sintering causes the oils and greases to
      vaporize but is insufficient to burn
      them, hence thesehydrocarbons deposit
      on air handling equipment,collect dust,
      and increase maintenance costs. Parti-
      culate air pollution control devices also

cannot tolerate these oils. For these reasons, iron oxides occurring in rolling operations are not widely used and must be disposed of as sludge.

The recovery process is a multistage process for reclaiming oil-free oxides and recovering the oils. Only around 3 percent of the original sludge is disposed of as waste. The sludge undergoes successive separation and washing cycles; oils and greases are washed from the particles and later separated from the wash solution; oxides are screened into coarse and fine grades.

The system produces a product with 31 percent higher iron content than Mesabi Range Fines and 15 percent higher than Marquette Range pellets. Costs of recovery are substantially lower than the cost of raw ores or pellets.

3.   Recovery of pickle liquors. An 85,000 ton/year stainless steel annealing-pickling line used nitrichydrofluoric acid in the pickling operations. Spring pickel baths were neutralized with lime, dewatered, and disposed of with loss of all the acid and pollution problems attending sludge disposal.

The recovery is accomplished by liquid-to-liquid extraction using an organic solvent, TBP, at a 75 percent concentration in kerosene. The solvent forms adducts with monobasic acids. The acids are stripped from the solvent with water and the solvent is reused. Sulfuric acid is added to increase acid yield. The metal sulfate by-products are neutralized before disposal.

The process recovers about 95 percent of the acid used; as a consequence of the regeneration process, overall acid usage has dropped by 47 percent.

4.   Sulfur and energy recovery. In this
case the company faced the need to
substantially reduce SO_2 levels and
odors from a sulfite mill. SO_2 was
emitted from a vent of the spent sulfite
liquor evaporator and from the blow tank
receiving gases from the digester at
intervals of two hours.

The company solved its problem by
ducting evaporator SO_2 emissions
directly to the mill's acid plant wet
scrubber and by adsorbing blow tank SO_2
emissions directly in the acid plant.
To deal with the blow tank emissions,
the company installed a system for
cooling and condensing blow tank off
gases ahead of absorption in an
absorption tower and for recovering heat
from the digester blow as an adjunct to
gas cooling. Recaptured SO_2 is
reutilized in the mill.

5.   Air pollution reduction. This is a
case where an inefficient system was
designed by the original manufacturer to
reduce particulate air pollution and to
recovery energy. The process involved
the handling of corrugated board scrap.
Waste materials were shredded and then
conveyed pneumatically to a roof-mounted
cyclone separator over a baling operation.
The system created dust, and plant air
was exhausted to the atmosphere. The
cyclone was sealed and equipped with a
dust filtering device. The captured
dust is conveyed automatically to the
baling system. Filtered air is returned
to the plant, and the incoming air
volume is adjusted to the operation of
the scrp handling system. Incoming
makeup air is now heated by spent steam
from corrugating machines, heat which
had been vented previously.

6.   Metal Recovery. A 180 tons per day
inicnerator was built to destroy plant
waste and sludge, produce energy, and
recover silver. The plant started
operations in 1976; it had been under
development for nearly 8 years.

The system receives waste from four manufacturing plants, a large office building, and an educational campus. Solid waste is shredded and separated into a light and heavy fraction. The light fraction is burned in a waterwill incinerator; combustion is augmented by burning fuel oil. Flue gases are used to dry sludge generated in a wastewater treatment plant before being burned. Steam generated in the incinerators is used in a turbine to produce electrical power.

Bottom ashes from the system and from the electrostatic precipitator are processed further to obtain silver. Silver is introduced into the system by waste film generated in the production operations.

As these cases illustrate, recovery, recycle, and reuse approaches take many technical forms--the use of new or modified devices and equipment; new processes that replace or augment existing processes; the modification of existing processes by a variety of techniques including changes in raw materials, the elimination of or change in process chemicals, capture and use of waste heat, recovery of solvents, cleansers, and a variety of raw materials, and the substitution of wastes for purchased fuels.

INDUSTRIAL WASTEWATER RECYCLE

A study conducted at Argonne National Laboratory [Kremek, 1981], identified the technology presently used to allow reuse of wastewater from some major industrial categories. The technology available to recover material from the metal plating and iron and steel industries provides good illustrations, as follow:

1.   Electroplating

Wastewater from plating processes comes from cleaning, surface preparation, plating, and related operations. The constituents in this wastewater include the basic material being finished as well as the components in the processing solutions. Predominant among the wastewater constituents are copper, chromium, nickel, zinc, lead, tin, cadmium, gold, silver, and platinum metals, as well as ions that occur from cleaning, surface preparation, or processing baths such as phosphates, chlorides, and various metal-complexing agents.

Wastewater from metal finishing processes comes from cleaning, pickling, anodizing, coating, etching and related operations. Predominant among the wastewater constituents are ions of copper, nickel, chromium, zinc, lead, tin, and cadmium, and ions that occur in cleaning, pickling, or processing baths such as phosphates, chlorides and various metals complexing agents.

Table 2.5 lists the variations in wastewater characteristics found in the electroplating industry. The table shows variation between plants as well as within each plant.

Table 2.5   Comparison of Raw Waste Streams
            from Common Metals Plating

Constituent	Range (mg/l)
Copper	0.032- 272.5
Nickel	0.019-2954
Chromium:	
Total	0.088- 525.9
Hexavalent	0.005- 334.5
Zinc	0.112- 252.0
Cyanide:	
Total	0.005- 150.0
Amenable to chlorination	0.003- 130.0
Fluoride	0.022- 141.7
Cadmium	0.007- 21.60
Lead	0.663- 25.39
Iron	0.410-1482
Tin	0.060- 103.4
Phosphorus	0.020- 144.0
Total suspended solids	0.100-9970

The results of analysis of the specific constituents of raw waste streams, from 50 metal finishing establishments are presented in Table 2.6 andTable 2.7.

Table 2.6   Composition of Raw Waste Streams from
             Anodizing (in mg/L)

Chromium, total	0.268- 79.20
Chromium, hexavalent	0.005- 5.000
Cyanide, total	0.005- 78.00
Cyanide, amenable to chlorination	0.004- 67.56
Phosphorus	0.176- 33.00
Total suspended solids	36.09 -924.0

Table 2.7   Composition of Raw Waste Streams from
             Coatings (in mg/L)

Chromium, total	0.190- 79.20
Chromium, hexavalent	0.005- 5.000
Zinc	0.138- 200.0
Cyanide, total	0.005- 126.0
Cyanide, amenable to chlorination	0.004- 67.56
Iron	0.410- 168.0
Tin	0.102- 6.569
Phosphorus	0.060- 53.30
Total suspended solids	19.12 -5275

Table 2.8 presents a summary of a survey showing that a substantial number of evaporation, ion
exchange, and reverse osmosis units are currently
used in a variety of recovery applications.

Table 2.8   Current Application of Leading Recovery
             Techniques for Electroplating and Metal
             Finishing

	Units in Operation		
Application	Evaporation	Ion Exchange	Reverse Osmosis
Chromium plating	158	50	——
Nickel plating	63	38	106
Copper plating	19	—	3
Zinc plating	7	—	3
Cadmium plating	68	—	—
Silver/gold plating	13	20	—
Brass/bronze plating	10	—	—
Other cyanide plating	6	—	—
Mixed plating wastes	——	11	6
Chromic acid etching	6	—	—
Other	16	2	1

2.   Iron and Steel

The production of steel involves three basic processes:

1.   Coal is converted to coke

2.   Coke is combined with iron bearing material and limestone and reduced in a blast furnace to form molten iron and

3.   Iron is then purified into steel in an open hearth, basic oxygen or electric arc furnace.

Two of the processes involved in this operation that can be documented to provide product recovery are as follows:

1.   Cokemaking

Cokemaking operations include by-product recovery and beehive facilities. Nearly all the metallurgical coke produced in the United States is made in by-product recovery coke ovens.

The by-product recovery process not only produces high quality coke for use as blast furnace or foundry fuels and carbon sources, but also provides a means of recovering valuable by-products of the distillation reactions. The volatiles are recovered from the gas stream and processed in a variety of ways to produce tars, light oils, phenolates, ammonium compounds, and napthalene (Table 2.9).

Table 2.9   Materials Recovered in By-Product Cokemaking Operations

Material	Extent of Recovery	Use
Crude coal tar	All plants	Resale and/or further processing
Crude light	Most plants	Resale and/or reuse
Ammonia and ammonium compounds	Most plants	Recirculation

Table 2.9 (continued)

Material	Extent of Recovery	Use
Phenol,pheno- lates, carbo- lates	Most plants steam strip-free ammonia from excess ammonia	Reuse on-site
	Half of these plants recover fixed ammonia	
Sulphur and sulphur com- pounds	About 1/3 of the plants desulphurize	Reuse on-site
Napthalene	About 70% of the plants	No information

2.   Acid Pickling

During the forming and finishing operations the steel is exposed to the atmosphere, causing an oxide scale to form on its surface. Before further processing, this scale must be removed as it inter- feres with the application of protective coatings to the steel and for cold rolling. An acid pick- ling solution is used to remove the scale. Regard- less of the type of acid used, sulfuric, hydrochlo- ric or a combination of these two, the spent pickel liquors (SPL) are highly contaminated. Approxi- mately 1.4 billion gallons of spent pickle liquor are generated annually: 500 million gallons of spent sulfuric acid; 800 million gallons of spent hydrochloric acid; and 74 million gallons of combi- nation pickling acids.

Following is an explanation of sulfuric and hydrochloric recovery and regeneration technolo- gies.

Sulfuric Acid Recovery - The most common treatment method for recovering valuable products from spent sulfuric acid is acid recovery by removing ferrous sulfate through crystallization. Spent pickle liquor, which is high in iron content, is pumped into a crystallizer, where the iron is precipitated (under refrigeration or vacuum) as ferrous sulfate heptahydrate crystals. As the crystals are formed, water is removed and the free acid content of the solution increases to a level where it is reusable in the pickling operation. The crystals are separated from the solution, and the recovered acid is pumped back to the pickling tank. The by-product ferrous sulfate heptahydrate is commercially marketable. The crystals are dried, bagged, and marketed, or sold in bulk quantities. Ferrous sulfate, commonly referred to as "copperas," is used in appreciable quantities in numerous industries, including the manufacture of inks, dyes, paints, and fertilizers. It is also used as a coagulant in water and wastewater treatment.

Hydrochloric Acid Regeneration - The only commercially proven technology to regenerate spent hydrochloric acid is through thermal decomposition. The spent pickle liquor contains free hydrochloric acid, ferrous chloride, and water. The liquor is heated to remove some of the water through evaporation and to concentrate the solution. The concentrated solution is the further heated to $925°$ to $1050°$ C ($1700°$ to $1920°$ F). At this temperature, water is evaporated and the ferrous chloride decomposes into iron oxide (ferric oxide, Fe_2O_3) and hydrogen chloride (HCl) gas. The iron oxide is separated and removed from the system. The hydrogen chloride gas is reabsorbed in water (sometimes rinsewater or scrubber water is used), to produce hydrochloric acid solution (generally from 15% to 21% HCl) which is reused in the pickling operation. There are several types of these "roaster" processes in operation.

REFERENCES

Kremer, F., Broomfield, B., "Recovery of Materials and Energy. From Industrial Wastewaters." Argonne National Laboratory, December 1981.

MRI Project No. RA-219-N, "Alternatives to Conventional Pollution Control", 1980.

CHAPTER 3

RESOURCE RECOVERY FROM
HAZARDOUS WASTE

INTRODUCTION

The generation of increasing amounts of hazardous and toxic wastes associated with the nation's rapid industrial expansion represents a problem of increasing public concern. The Environmental Protection Agency has recently presented statistics indicating that 10-15 percent of the 344 million metric tons of wet industrial wastes that are produced each year can be classified as being hazardous [US EPA, 1974].

Hazardous wastes have been variously defined. The Resource Conservation and Recovery Act of 1976 (RCRA) defines a hazardous waste in Section 1004(5) as:

- A solid waste, or combination of solid wastes which, because of its quantity, concentration, or physical, chemical or infectious characteristics may –

    - cause or significantly contribute to an increase in mortality or an increase in serious irreversible, or incapacitating reversible illness, or

    - pose a substantial present or potential hazard to human health or the environment when improperly treated, stored, transported, or disposed of, or otherwise managed.

More recently EPA has attempted to identify hazardous wastes in terms of certain characteristics as [US EPA, 1974]:

- ignitability
- corrosivity
- reactivity
- toxicity

Resource recovery is attractive both for economic and environmental reasons. Recovery of waste energy or material value is becoming an increasingly viable option, as ultimate disposal options become more strictly regulated and expensive. The market for recovery has begun to develop an industry, including centralized commercial processing and recovery facilities, and industrial waste exchanges.

Resource recovery typically involves recycle and reuse, either with or without pretreatment for purification before reuse. Each increment of hazardous waste which is recycled has value, and represents an increment of material not requiring detoxification and/or ultimate disposal.

The State of California has recently reported five alternatives to landfill which are to be encouraged by regulatory and other efforts [Patterson and Haas, 1982]. In order of preference these are: source reduction by modifying industrial processes to generate less waste; waste recycling and resource recovery; physical, chemical, or biological treatment that renders the waste innocuous; high-temperature incineration for many organic compounds; and solidification or stabilization methods that chemically fix or encapsulate the wastes so that they are less mobile in the environment. Since 1977, the State of California has had an active program to investigate the feasibility of recycling hazardous wastes and to develop techniques to encourage such reuse in California. This program includes research and development, and technical liaison through a combined clearinghouse-consultation approach with industry, and has resulted in a successful and growing trend of recycle and recovery. On the basis of experience to date, California has identified five broad recycling categories of hazardous wastes;

Type I   Unused commerical chemicals in packages.

Type II	Process wastes that are economically and technologically feasible for recycle without prior treatment.
Type III	Process wastes that are economically and technological feasible to recycle if pretreated and purified.
Type IV	Process wastes that are not presently economically and technologically feasible to recycle.
Type V	Wastes that are undesirable to recycle.

Type IV wastes may shift to the Type II or III category as the experience and data base for recycle expands. Type V wastes include those extremely hazardous substances (carcinogens, pesticides) that must be destroyed due to bans or restrictions on their use or reuse.

RESOURCE RECOVERY

Before specific processes and alternatives for resource recovery are discussed, it is necessary to describe broad strategies for the recovery of economic value from hazardous waste processing. In general, there exist four broad paths whereby recovery of some value from waste may occur:

- Direct recycle for primary (generator) use

- Use by a second industry as a raw material

- Energy recovery

- Utilization in pollution control systems

A given hazardous waste stream may be a potential candidate for recovery by more than one of the above routes; however consideration of the benefits to be derived will provide guidance in determining needed pretreatment methods to be used.

Direct Recycle

One route for reuse of a hazardous waste is the processing of the waste stream to recover materials of value to the industry which generated the waste. For larger generators, strong economic incentives to implement direct recycle options exists. However, for smaller generators, the cost of recovery may negate any value to be recovered; here the possibility exists for collection and recovery by a contract processor, or establishment of a centralized cooperative facility.

Some instances of direct recycle, elaborated below, include chromic acid recovery from spent plating bath solutions, and organic solvent recovery from degreasing operations.

Recycle to a Second Industry

By definition, the wastes disposed of by an industry have no intrinsic value to that industry so, in many cases, direct recycle is not a viable option. However, the waste from a given industry may contain material of value to another industry, and may represent a competitive source of supply of that material, with or without intermediate purification and/or enrichment.

Example possibilities for secondary recovery include solvent recovery, recovery of phenols from coking wastes, sulfur recovery from stack gas cleanup, and metal and/or acid recovery from various pickle liquors. Industrial waste exchanges often perform a secondary recovery function, by accepting wastes from one site, and providing it as a process chemical to another.

Energy Recovery

Many hazardous wastes, particularly those containing organic matter, have sufficient energy value to enable recovery of energy to be economically viable. The most common implementation of this strategy is the use of spent solvents or waste oils for steam generation. An additional and attractive option would appear to be the use of waste chlorinated solvents as alternative fuel sources in cement manufacture. Energy recovery may occur at the site of the generator, or at a secondary site.

Recovery for Pollution Control

In some cases, the value of a given waste stream may be realized by using the waste stream in the control of other pollutants, in place of more expensive alternative treatment reagents. The most common use of this strategy is the neutralization of waste by mixing an acidic and an alkaline waste stream. Another promising strategy is the use of waste pickle liquor for hydrogen sulfide or phosphorous control in wastewater treatment plants.

UNIT PROCESSES USED IN RESOURCE RECOVERY

The treatment which a specific hazardous waste receives prior to resource recovery may involve a number of discrete unit processes, which depend upon the nature of the material, the nature of the desired end-product and the type and extent of contamination. Many other factors will influence the selection of processes that are chosen, such as economics, geographical considerations, and federal, state and local environmental regulations. Specific examples are discussed in a later section, with individual wastes considered.

EXAMPLE RESOURCE RECOVERY OPTIONS

Resource recovery options exist for many of the types of hazardous wastes, and will be discussed in this section.

Recovery from Acidic Wastes

As indicated in Chapter 1, it is estimated that about 35 percent of all liquid wastes contain acids. In the case of acid recovery from relatively uncontaminated solutions, the purification and recovery of the acid is technically feasible. However, the economics strongly depend on the acid concentration, degree and type of acid contamination, and local factors, such as transportation costs.

In the case of nitric acid wastes, recovery may be effected by steam distillation. The recovered product, after condensation, consists of a mixture of 68 percent acid and 32 percent water, which is the constant boiling point composition. This product may then find numerous industrial uses. The waste from the recovery operations, consisting of still bottoms, would be neutralized and, where necessary, given further treatment, and any residue disposed of in a secure landfill [Ottinger, et al., undated].

In the case of wastes containing sulfuric acid, recovery can be effected if the following requirements are met:

- The recovery operation is near an existing sulfuric acid plant.
- Waste contains more than 70 percent acid.
- Waste volume is greater than 50 tons/day.
- Organic impurities are low.
- Inorganic impurities are low.

The recovery process involves the use of heat to decompose the acid into sulfur dioxide, which evolves as a gas. The gas stream is then collected and piped to the sulfuric acid manufacturing operation for use as a raw material [Ottinger, et al., undated].

Acid Solutions Containing Metals

Acidic solutions containing metals arise from metal finishing operations, from electroplating operations, both as spent plating solutions and as rinsewaters, and from the steel industrial processes, among others. There may be a potential for recovery of either or both the acid and the metal contained in the waste.

In the case of waste pickle liquor, it is possible to use the liquor directly in pollution control. Ferrous iron salts, which may constitute as much as 15 percent of the waste, have been useful for the removal of phosphorus from wastewater effluents by precipitation [Ottinger, et al., undated].

Recovery of metals from other wastewaters may be effected by various precipitation or ion exchange processes. These have generally been designed or discussed on a case-by-case basis, with the principal exception of chromium recovery. Weber (1972) has outlined a three-bed process for the treatment of chrome plating wastes, involving recovery of treated water, strong (4-6 percent) chromic acid

solution, and separated metal sulfates in the regenerant waste stream.

     In the case of most other metals, the initial
step in recovery from an acidic waste stream would
appear to involve neutralization and precipitation.
The resulting sludge can then, in some cases, be
treated for metal recovery. One process proposed
involves the leaching of these sludges with sulfuric acid, and filtration, to remove solid matter,
such as calcium sulfate. The filtrate may then be
neutralized to pH 3, with an alkali, to precipitate
any iron hydroxide, for recovery. Remaining
filtrates can be electrolysed for recovery of
metallic nickel copper. The supernate from the
electrolysis operation can then be neutralized with
lime to a pH of 9, and cadmium, chromium and zinc
recovered from the precipitate. One economic analysis of this process configuration indicated that,
for a plant treating 50 tons/day of metal sludge, a
net income, exclusive of costs associated with
ultimate disposal of residuals, of $572/day was
possible [Battelle Memorial Institute, 1974].

     In certain cases reductive precipitation of
metals from waste streams may be practicable. In
the case of chromium, reduction of hexavalent
chromium to trivalent chromium, followed by alkaline precipitation, produced a sludge containing
chromium hydroxide. It has been proposed that this
sludge may be treated with sulfuric acid, for the
recovery of chromium sulfate, and subsequent reuse
in the leather tanning industry [Ottinger, et al.,
1973]. In view of the geographic concentration of
leather tanning in New York, and the proposal for a
central treatment facility for this industry
[Iannotti, et al., 1979], implementation of such a
chrome recovery step may be appropriate.

     For the recovery of nickel, copper, and silver
from wastewater, reductive precipitation has also
been proposed. After silver solutions are contacted with zinc dust, metallic silver can be recovered [Battelle Memorial Institute, 1974].
Treatment of wastes containing nickel or copper
with iron filings at a pH in excess of 3.5 may also
result in the recovery of metallic nickel or copper
[Battelle Memorial Institute, 1974].

Acid Solutions Containing Organics

Many wastes fall in this category, ranging from minuscule flows to very large flows and varying from low concentrations of both acids and organics, to very high concentrations of either or both components. The organics may be present either in emulsified form or dissolved form.

Resource recovery may be directed in one of two directions for waste streams containing substantial quantities of components.

- neutralization of acid and recovery of organics
- removal of trace organics and recovery of acids

Neutralization of acid wastes containing emulsified organics often causes the emulsion to break and allow the separation of a substantial fraction of the organic material which can be removed by skimming or decantation. If the quantity of organics warrants recovery, it may be fractionated and purified by distillation. Otherwise, the organic layer can be removed and incinerated. Any residual dissolved organics may be removed and incinerated. Any residual dissolved organics may be removed by one of several techniques such as adsorption, flotation or extraction, or one of the biological treatment methods. Recovery of grease and oil from acid wastes is often feasible, e.g., acid waste streams from refineries, electroplating, or metal finishing operations [Hall, 1978; Tabakin, 1978; Luthey, 1978; Steiner, 1978; Humenick, 1978; Levin, 1981].

If the quantity of organics in acid wastes is small they may be removed by techniques such as adsorption, flotation or extraction. The remaining acid, if in attractive quantity, can be removed and recovered or possibly reused by techniques discussed earlier in this section. It should be noted that increased attention has recently been given to the recovery of acids from pickle liquor [Wadhawan, 1978].

Acid Sludges

Acid sludges come from a wide variety of industrial processes and vary widely in quantity and composition. They often are highly viscous which makes further processing quite difficult. It, therefore, is unlikely that any great potential exists for recovery except in special situations which must be evaluated on a case-to-case basis.

One exception to the above situation is the case of acid sludges from electroplating operations. These sludges may contain metals together with high concentrations of acids which make the waste hazardous. Recovery of the metals may be accomplished by extraction with suitable solvents or acids and subsequent recovery of the metals by ion exchange or neutralization, precipitations and filtration. Thus, recovery, treatment and disposal can be incorporated in the same process [Gurnham, 1965; Nemerow, 1978; Leonard, 1971].

Acid Gases

The potential for recovery or reuse is small except in some special cases. For instance flue gas is sometimes used to neutralize alkaline liquid wastes, utilizing the acidic components of SO_2 and CO_2 [Nemerow, 1978; Steele, 1954; Cana, 1959; Fisco, 1960]. Some instances of HCl recovery by absorption and concentration have also been reported [Nemerow, 1978; Besselievre, 1969; Gurnham, 1955].

Resource Recovery from Alkaline Wastes

Alkaline wastes may be generated in a variety of industries, including electroplating and textiles. Some of these wastes may contain cyanides and metals, as in electroplating; other wastes may contain high concentrations or organics and/or solids, as in textile manufacture. Resource recovery strategies may include recovery of the alkali itself, or recovery of the other materials contained therein. In addition, the possible use of any alkaline material as a scrubber feed should be mentioned, for possible use in the removal of acid gases from incinerator stack gases.

Alkaline Solutions Containing Metals

The major class of wastes of this group is that arising from electroplating wastes containing metal cyanides. This type of waste is widespread in many major industrial centers. A potential

exists for resource recovery, and various recovery
strategies have been described.

Three possibilities for recovery have been
described. A simple evaporative distillation to
recover water, and concentrated metal cyanide solution can be carried out to permit reuse at the
point of generation. [Ottinger, et al., 1973]. A
second strategy for metal recovery alone involves
the addition of metallic iron to the waste, to all
the precipitation of elemental copper and nickel
from the solution [Ottinger, et al., 1973]. The
residual will contain cyanide, which must be further processed prior to ultimate disposal.

As a third approach, both the metal and cyanide may be recovered if an ion exchange process is
employed. In this case, two different beds are
used--an anion exchanger and a cation exchanger,
and the waste is passed through the two columns in
series. In the cation exchanger, regeneration
yields a metal solution, typically metal chloride,
if brine is used as a regenerant. In the anion
exchanger, use of either caustic or brine as a
regenerant. In the anion exchanger, use of either
caustic or brine as a regenerant will allow recovery of sodium cyanide. The resulting metal
chloride stream and sodium cyanide stream may then
be used in the plating process, or sold for a
secondary use [Ottinger, et al., 1973; Booz Allen
Research Inc., 1973].

Alkaline Solutions Containing Organics

If the organic material present in these
wastes is at a sufficiently high concentration, and
of a nature where recovery is desirable, then alkaline solutions containing organics may be treated
for recovery of the organic fraction according to
procedures described in a later section. For
several waste streams, however, the nature of the
organic contaminant, or its concentration, are such
that organic recovery is either not practicable or
economic. In these cases, recovery of the alkali
might be investigated.

As an example of this possibility, wastes from
textile processing may consist of a strong caustic
solution, routinely contaminated with hemicellu-
lose. In a full scale installation, use of dialy-
sis for the recovery of a 9-10 percent caustic

solution, free of organic impurities, has been
found feasible, and has resulted in a recovery
of 200 tons/yr of sodium hydroxide [Iannotti, et
al., 1979]. It would be anticipated that this
technique can be applied to strong alkaline solutions containing organic contaminants, provided
that other inorganic contaminants and suspended
solids concentrations were relatively low.

Alkaline Sludges

Alkaline sludges may arise in a number of
industrial processes. The most common type of
alkaline sludges for which resource recovery processes have been proposed is the sludge resulting
from chloralkali manufacture. These wastes may
contain lead where the diaphragm cell process is
used, and mercury from electrode contamination and
brine impurities.

It has been proposed that lead recovery from
waste sludges of the diaphragm cell process could
be carried out using smelting. The sludge to be
treated is dewatered, using coagulants, and filtration, to 25 percent solids. The dewatered sludge
can then be blended with lime, silica, and coke,
and smelted in a reducing atmosphere at $1000-1040^{\circ}C$
for lead recovery. Using this process, air pollution controls on the smelter would be necessary
[Shaver, et al., 1975]. The estimated cost of this
process was $0.81/metric ton of chlorine produced.

Two mercury recovery processes for chloralkali
sludges have been proposed, and are currently in
use in industrial applications. Georgia-Pacific in
Bellingham, WA, has been reported to use a dewatering process, followed by roasting to recover
mercury. The mercury vapor is recovered from the
combustion gases by condensation and demisting.
The data from the full scale installation permits a
cost estimation, including recovery of mercury, of
$2.50-2.90/metric ton of chlorine produced [Shaver
et al., 1975].

Mercury recovery from a similar waste by a
non-thermal process is practiced at the BASF
Wyandotte plant, Fort Edwards, WI. The brine purification muds are acidified to pH 2 with waste
concentrated sulfuric acid, and calcium carbonate
is allowed to precipitate. The mixture is then
treated with sodium hypochlorite at pH 6-7 to

dissolve the mercury from the sludge. Vacuum filtration to remove the remaining low-mercury solids
for disposal, and collection of the mercury-laden
filtrate is then performed. The filtrate is then
combined with other mercury laden wastes, and
treated with sodium bisulfite, sulfuric acid and
sodium hydrosulfide for about 16 hours to precipitate mercuric sulfide. The mercuric sulfide can be
recovered as a filter cake by leaf filtration. The
effluent can then be discharged, after monitoring
for mercury levels. The mercuric sulfide may be
used off-site, or dissolved in sodium hypochlorite
and the solution recycled to the main plant brine
circuit. Operating data from full scale installation indicate a net cost of $6.90/metric ton of
chlorine produced [Shaver et al., 1975].

Resource Recovery from Other Inorganic Wastes

Various waste streams may be generated that
contain hazardous inorganic materials such as salt
solutions, solid salts, metals and non-metals. A
wide range of materials and concentrations are
encountered, and the possibility of recovery or
reuse must be evaluated for individual situations.

Salt Solutions

There does not appear to be a high potential
for recovery from wastes of this type. Some process solutions may contain hazardous salts in sufficient concentrations to warrant recovery, such as
fluorides from smelting operations [Gurnham, 1965].
Various concentration techniques such as evaporation may be necessary as pretreatment, and are
usually expensive because of investment costs and
high heat costs. Reuse of waste salt solutions can
sometimes be practiced, either within the same
process from which waste originated or in neigh-
boring processes or other industries.

Solid Salts

Various metal-finishing industries discharge
batches of salts used as molten salt baths for
heat-treating. These baths often contain cyanides,
fluorides, and other salts with toxic properties.
Recovery from these mixtures is possible by dissolving, filtering and recrystallizing the salts.
Such processing is, however, unlikely to be cost
effective [Ottinger et al., 1973; Shaver et al.,

1975; Genser et al., 1977].

Metals

The electrolytic refining of copper results in the generation of an anode mud which contains toxic metals such as arsenic, antimony, nickel, selenium and tellurium. The concentrations of these materials may be high enough to justify extraction and recovery of these materials by ion exchange or electrolysis [Gurnham, 1965].

Some metal refineries and other industries generate metallic dusts which are usually water scrubbed for removal of entrained solids. Recovery from the scrubber water by filtration is practiced [Leonard et al., 1977].

Wastes discharged from some chlorine-alkali electrolytic cells often contain substantial amounts of mercury. The mercury can and should be recovered by controlled incineration or retorting followed by condensation of the mercury combined with vapor cleanup with mist eliminators, adsorption and activated carbon or scrubbing with hypochlorite solutions [Ottinger et al., 1973; Shaver et al., 1975; Booz Allen Research Institute, 1973].

Lead is frequently recovered from storage batteries and other waste materials. Carefully controlled smelting or burning techniques have been used to recover the lead. The value of lead makes its recovery attractive [Gurnham, 1965; Battelle Memorial Institute, 1974].

Non-Metals

A potential exists for the recovery and reuse of sulfur from a number of common hazardous waste streams. Many processes found in such industries as the coke and steel industry, tanneries and various chemical industries generate sulfur compounds. These materials often appear in alkaline scrubbing wastes or other waste solutions. The sulfur compounds can often be concentrated by steam stripping, converted to hydrogen sulfide and then reduced to elemental sulfur by the Claus Process or one of its modifications [Ottinger et al., 1973].

Resource Recovery from Concentrated Organic Liquids

The potential for resource recovery from concentrated organic liquid wastes arises from the
economic, material and energy values resulting from
the presence of certain specific organic materials
(e.g., solvents) within the waste stream. For
lightly contaminated material, resource recovery is
relatively easier than for heavily contaminated
waste. However, in general, all strategies embody
some degree of purification, which results in both
recovery of a more usable product, and partial
detoxification and volume reduction of the residual
waste.

Lightly Contaminated Wastes

In a wide variety of industries, recovery of
solvents from lightly contaminated wastes is feasible, and currently practiced. It has been estimated that some pharmaceutical companies recycle
over 99 percent of the hazardous solvents which
they use [Booz Allen Research Institute, 1973]. In
the electronic components industry, surveys indicate that over half of both the halogenated and
non-halogenated solvents are segregated for reclamation [Peters et al., 1977]. A wide variety of
solvents and other organics may thus be reused,
including the halogenated methanes and ethanes
[FEA, 1976].

Techniques used for solvent and organic recovery vary with the nature of the waste and the
product desired. In some cases, relatively clean
solvents may be wasted, and simple repacking for
use by other commercial users is adequate [Peters
et al., 1977]. More commonly, distillation and
recondensation are used to recover desired solvent
[Peters et al. 1977, FEA, 1976]. Several commercial reclamation services employ extensive distillation or fractional distillation processes for
solvent recovery.

A detailed process design has been published,
involving multiple distillation and condensation
steps, which is claimed suitable for separation and
recovery of the following organic solvents: methylene chloride, chloroform, methyl chloride, ethylene dichloride, trichloroethylene, perchloroethylene, and o-dichlorobenzene [Battelle Memorial
Institute, 1974]. It must be emphasized that in

any recovery strategy such as this, a residual will
always be present, typically as a still bottom, and
must receive adequate consideration for proper
disposal.

Heavily Contaminated Wastes

In the case of heavily contaminated wastes,
recovery of organic materials may be effected in a
similar manner to that for lightly contaminated
materials, especially in cases where there is a
significant difference in boiling point between the
desired material and the contaminants. While the
potential needs to be evaluated on a case-by-case
basis, several examples of promising recovery
techniques have been reported.

In the wool industry, scouring and degreasing
operations may result in a solvent heavily contaminated with lanolin and other greases. The solvent may be distilled and recycled within the
plant; the still bottoms, consisting primarily of
lanolin and grease, may also be recovered, if a
satisfactory market is available [Booz Allen
Research Inc., 1973].

In the petroleum industry, wastes from process
condensate or steam stripper condensate or bottoms
may be heavily contaminated with phenol. A proprietary solvent extraction process, the PHENEX
process, has been proposed for recovery of phenol
and removal of this material from the liquid stream
[Booz Allen Research Inc., 1973].

In a number of industries, heavy ends or still
bottoms are produced from distillation or other
processes. Typically, in the organic chemical
industries, liquid or semisolid wastes may result
from the manufacturing process. Again, resource
recovery from wastes of this nature may require a
case-by-case analysis. However, two case studies
are given below to illustrate the potential for
resource recovery.

In the manufacture of chloromethane solvents,
a waste stream is produced consisting of hexachloro-
benzene, hexachlorobutadiene, and miscellaneous
materials. An initial distillation of this waste
yields a distillate which may be chlorinated at
high temperature and pressure in a nickel tube, to
yield carbon tetrachloride. After quenching and

partial cooling and depressurization, the high
boiling compounds may be distilled off and recycled
back to the chlorine contactor. The remaining
material may be separated into chlorine and carbon
tetrachloride by distillation. The carbon tetrachloride may be further purified and marketed with
the production from the main plant. Economics of
this process indicate favorable costs of about 1-2
percent of the total product value [Genser et al.,
1977].

In the manufacture of perchloroethylene, a
waste stream is produced which consists primarily
of hexachlorbutadiene. By stripping the volatile
material from this waste, recycling this to the
main plant, and distillation of the remainder,
hexachlorobutadiene can be produced. Including
costs for disposal of the still bottoms from the
recovery operation, economic analysis indicates the
potential for a net profit from the treatment/
recovery operation [Genser et al., 1977]

Resource Recovery from Dilute Aqueous Solutions of Organics

There does not appear to be any significant
near term potential for resource recovery from
dilute organic aqueous solutions. One possible
exception to this statement may be in the case of
highly specialized waste streams, such as those
from the pharmaceutical industry, which might contain low concentrations of relatively valuable
material. The major basis for this pessimism lies
in the relatively high costs of probable separation
processes which might be used to purify and concentrate the dilute organic contaminants into a usable
product.

A possible recovery strategy for this group of
wastes currently forseeable is the recovery of
energy from this material when activated carbon is
used in removal. In this case, during the thermal
regeneration of carbon, the process might be operated to provide some beneficial use of waste heat.
It would not appear, however, that this option will
permit a net production of energy in the GAC process.

Resource Recovery from Organic Solids

Although the major aims of hazardous waste

disposal of organic solids are normally detoxification and volume reduction, opportunities exist for
resource recovery. The general strategies used are
energy recovery, direct recycling as a raw material
in another industry or use of a pyrolytic system to
generate liquid or gaseous fuels. Energy recovery
from thermal processing is considered in a later
section.

Contaminated organic solids (or liquids) which
cannot be recycled must ultimately be disposed of
by incineration. In some cases, recovery of the
contaminate will result in a usable product. For
example, hydrogen chloride is produced during incineration of chlorinated hydrocarbons. To meet air
pollution regulations, this gas must be collected.
Since hydrogen chloride or hydrochloric acid has
economic value, its presence then becomes a credit
rather than a liability [Novak, 1970; Ross, 1977;
Santoleri, 1973].

Salts and Pure Organic Compounds

Although the volume of waste in this
category is, in many locations, very small,
salts and pure organic compounds are among the
best candidates for direct recycling [Tabakin
et al., 1978]. Wastes from one industry can
sometimes be used directly as a raw material
for another industry. This exchange can only
take place when a mechanism exists for easy
transfer. Hazardous waste clearinghouses can
provide this avenue for exchange [US EPA,
1976; Terry et al., 1976].

Tars and Residues

Tars and residues generally have little
value for resource recovery, other than their
heat content. Energy recovery after thermal
processing is possible if the heat content of
the waste is high enough. However, with
increased energy costs the fuel quality of
residues has been steadily decreasing [Novak,
1970].

Sludges

Since most hazardous organic sludges
will commonly require incineration, several
possible methods of resource recovery might be

feasible other than direct recovery for reuse.

Recovery of inorganic contaminants after
incineration from the scrubber water or ash
is possible. Metal oxides have economic
value, and can be recovered by extraction and
ion exchange. These compounds can be
reprocessed to recover the metals. Recovery
of halogens from the scrubber water is another
possible direction. For example, brominated
tars and sludges, when incinerated, release
free bromine. Recovery of the relatively
valuable bromine, rather than its complete
disposal, seems practicable [Novak, 1970;
Sebastian, 1975; Folks et al., 1975;
Santoleri, 1973; Ross, 1977; Hitchcock, 1979].

Resource Recovery from Organic Gases and Vapors

Organic gases and vapors will usually be incinerated or recycled at the point of production.
Recycling is accomplished by solvent stripping,
extraction or condensation. Tank gases have the
possibility of direct recycling as a raw material
for another industry. Unwanted gases and organic
vapors produced during disposal should be incinerated and can be used for energy recovery.

ENERGY RECOVERY

The recovery of energy and/or heating value
from certain organic hazardous wastes is deserving
of special consideration. In general, few incineration systems for toxic waste decomposition have
been designed to recover energy since auxiliary
fuel is almost always needed to ensure complete
combustion with minimization of undesirable atmospheric emissions. However, knowledge of detailed
feed composition and careful blending of wastes fed
to an incinerator will reduce the auxiliary fuel
needs, and thus reduce direct energy costs associated with incineration.

There are two basic manners in which energy
recovery may be effected in thermal processing:
heat recovery, and recovery of fuel value. Recovered heat may be used to preheat combustion air,
or in steam generation for heat or power production. Heat exchangers are generally tubular,
plate, or regenerative, and used for heat transfer
between the exit gases of a combustion process and

a working fluid. If air is the working fluid across a heat exchanger, it can be used as input to the combustion chamber to minimize fuel use. If the working fluid is water, the resulting steam or hot water may be sold or used in plant processes.

The second major type of energy recovery is utilization of the fuel value directly. Under pyrolytic conditions, a combustible liquid or gas may be produced from hazardous waste, which can be sold or used elsewhere [Novak, 1970; Ross, 1977; Boucher et al., 1977; Folks et al., 1975; Sebastian, 1975; Hitchcock, 1979]. An interesting opportunity for conservation of the fuel value of hazardous waste is the utilization of these materials as feed for rotary kilns in cement manufacture. Test burns of chlorinated hazardous wastes at the St. Lawrence Cement facility in Ontario showed a 99.98 percent destruction of PCB's and a 99.99 percent destruction of chlorinated organics [US EPA, 1975]. In this process, halogen gases are scrubbed into the cement product, and the fuel value of the hazardous waste directly reduces the consumption of fossil fuel. A typical cement plant may require 10^{10} BTU of fuel per day, and initial indications suggest that substitution of up to 15 percent of this fuel value with chemical waste is possible [Alpha Portland Industries, personal communication]. Certain industries produce solvent laden air streams, which must normally be treated to control air emissions. One alternative to treatment and production of a solvent waste for disposal is to directly use the solvent laden air as the air feed to a boiler. The fuel value of the solvent is then recovered in the boiler.

In energy recovery from hazardous wastes, as in incineration of chemical wastes in general, the issues of materials deterioration remain relatively poorly understood. Operating problems such as corrosion, erosion, plugging, fouling, and refractory decomposition can decrease useful life of components. In addition, any utilization of energy in outside industries necessitates the production of a consistent amount of steam and/or fuel, and thus a relatively constant throughput of hazardous waste, both in terms of quality and quantity, must be maintained.

FACILITATION OF RECOVERY

Industrial materials are perceived as special or hazardous wastes requiring disposal when, in the judgment of the waste generator, there is insufficient economic value associated with the waste to warrant alternative modes of utilization. This judgment may be predicated upon institutional regulations, or a lack of awarenesss of a potential secondary industrial market for the material. Facilitation of recovery thus must consider both institutional options and strategies to encourage alternatives to ultimate disposal, and methods to identify secondary recovery and reuse markets. This latter aspect is usually addressed through waste exchangers and information clearinghouses.

Institutional Options

The National Conference of State Legislatures has recently completed a survey and analysis of state policy options to encourage alternatives to land disposal of hazardous wastes [Patterson and Haas, 1982]. This analysis identified several state options to encourage the use, reuse, reclamation, and recycling of hazardous waste. The states can encourage such alternative to land disposal by providing financial, legal and institutional incentives and disincentives.

A variety of financial strategies have been described for encouraging alternative methods for management of hazardous waste. They include the use of fee structures, tax incentives and bonds. Fees can be structured to discourage undesirable disposal options such as land disposal, by making them higher for those generators who do not utilize use and reuse methods. States can encourage alternatives by setting lower fees for permits and licenses required of facilities that treat or recover hazardous waste. Tax incentives offer a positive inducement for business and industry to adopt alternative waste management methods. They include property and equipment tax exemptions, corporate income tax exemptions, accelerated depreciation schedules and sales, and use and excise tax exemptions for facilities which treat and recycle hazardous waste. Industrial revenue bonds may also be authorized to finance resource recovery facilities.

Illinois, for example, currently levies a uniform fee on all owners and operators of hazardous waste sites. Some other states structure fees to be higher for, or to be only levied on, those generators who do not utilize reuse and reduction methods. Kansas authorize its Department of Health and Environment, and Tennessee its Solid Waste Control Board, to establish a schedule of fees based on the degree of hazard and costs for treatment and disposal. In effect, these approaches allow the state to discourage high hazard waste disposal by scheduling higher fees for these facilities. Indiana places a tax of $1.50/ton on hazardous waste disposal, but it appears that this tax does not apply to resource recovery. In Ohio, those entities that detoxify or incinerate hazardous waste are not required to pay the fee which supports the state's special hazardous waste account. Such an approach places the financial burden on those facilities which would have the greatest tendency to call on the resources of the fund.

Recent Kentucky legislation authorizes the Department of Natural Resources and Environmental Protection to collect a hazardous waste tax. The annual hazardous waste management assessment is determined according to the quantity and volume of hazardous waste generated. The assessment for on-site treatment and disposal is one-half the amount for off-site treatment and disposal. On-site resource recovery and treatment facilities are exempt from any assessment, unless the process involves the landfilling of hazardous waste. The legislative intent of the assessment is to reduce the amount of waste generated, promote alternative to landfilling, and encourage on-site as opposed to off-site management. The assessment on generators of waste destined for long-term containment without prior treatment is significantly higher than for generators of waste destined for treatment. This approach rewards those generators who provide pretreatment of their waste. In addition, the Department has developed a schedule of fees for the costs of processing applications for permits and exemptions that is lower for recyclers of hazardous waste. Maine's proposed three-tiered fee structure is similar to Kentucky's in that license fees, renewal fees and taxes on generators of hazardous waste will be lower for resource recovery activity.

Florida offers an alternative approach.
Rather than raising the fees for generators who do
not treat their waste, facilities which render
waste non-hazardous are exempt from the four percent excise tax. The tax is "to be paid by each
generator of hazardous waste in the state...for the
privilege of generating hazardous waste." The tax
is levied at four percent of the price of disposing, storing, or treating hazardous wate. Moreover, the tax is levied in addition to all other
taxes imposed upon or paid by the generator. In
addition to exempting treatment facilities which
render the waste non-hazardous, the law exempts on-
site generation and disposal. As in Kentucky,
treatment and/or on-site facilities are encouraged.
Along with money collected from permitting fees,
fines, and appropriations, the money collected from
the excise tax is used to support the state's
Hazardous Waste Management Trust Fund. Table 3.1
is a summary of state fees on hazardous wastes
disposal.

Similar to fee structures in that they may be
used to encourage alternatives to land disposal,
tax incentives offer a positive inducement to business and industry to adopt alternative waste management methods. While fees may be scheduled as a
disincentive to certain practices, taxes can be
structured to reward those businesses and industries that engage in treatment or recovery of hazardous waste.

There are several types of tax incentives that
can be applied specifically to hazardous waste
treatment. Tax incentives provided for solid waste
facilities or pollution control equipment may also
be applied to hazardous waste management with
amendment of statutory language. The range of tax
incentives includes property tax exemptions,
"breaks" on equipment taxes, corporate income tax
exemptions, and sales, use and excise tax exemptions. An amendment (S.169) to the Federal Internal Revenue Code, and intended to provide tax-
exempt financing for process changes which prevent
the creation of pollution (e.g. source reduction
and resource recovery) was introduced in the 97th
Congress (Congressional Record, 1981). This proposed amendment specifically cites examples of
pollution prevention and resource recovery to be
covered under the amended Code.

Table 3.1   Summary of Selected State Hazardous
             Waste Disposal Fees [Bulanowski, et
             al., 1981]

State	Volume Basis	Fee, $
Alabama	gallon	0.036
	ton	5.00
Florida	see Footnote (1)	****
Illinois	gallon	0.01
	cubic yard	2.02
Indiana	ton	1.50
Iowa	see Footnote (2)	****
Kansas	cubic yard	0.25
Kentucky	gallon (1981)	0.02
	gallon (1983)	0.05
Maine	gallon (on-site)	0.12
	(off-site)	0.15
Missouri	see Footnote (3)	****
New Jersey	see Footnote (4)	****
Ohio	see Footnote (5)	****

(1) 4 percent tax on charge for disposing of waste
(2) 2 percent surcharge tax
(3) 2 percent tax on gross charges and fees
(4) 5 percent tax on gross receipts of disposal
     facility
(5) 4 percent tax on gross charges

Although almost all states authorize the is-
suance of industrial revenue bonds, only five
states specifically address their use in resource
recovery. These states are North Carolina,
Florida, Illinois, Mississippi, and Georgia. In
Illinois, facilities for which such funds are
available include those engaged in "reducing, controlling or preventing pollution...(or those that)
reduce the volume or composition of hazardous waste
by changing or replacing manufacturing equipment or
processes...recycle hazardous waste, or recovery
resources from hazardous waste."

Legal and institutional policies available to
encourage alternatives to land disposal range from
legislative and regulatory incentives and disincentives to the establishment of state research and
development programs. In Illinois, a portion of
the state hazardous waste disposal fee is allocated
under the enabling legislation to a research and
development fund. Legal options consist of
excluding recycled and reused materials from regulatory programs; eliminating permit requirements
for resource recovery facilities or on-site hazardous waste systems; expediting and the permitting
process (i.e. fast-track permitting) for recycling
facilities; restricting the burial of certain hazardous waste when it is feasible that it be treated
or recycled; and lessening the liability standards
for alternative management technologies.

There are many approaches to land disposal
restrictions. Perhaps the strongest position that
can be taken is an outright ban on land burial.
New York's Department of Environmental Conservation
has announced that they are currently writing regulations to "ban landfilling of environmentally
persistent and highly mobile chemical wastes..."
Further, New York has denied permits on the grounds
that they failed to adequately provide for technologies that offer alternative to land burial.
Another approach to banning land disposal of hazardous water is to require neturalization, detoxification, solidification or encapsulation of the
waste prior to land disposal.

Effective January, 1983, the State of
California banned the land disposal of six
categories of hazardous wastes which represent
about 40 percent of all hazardous wastes which
otherwise would be deposited in landfills. The

banned wastes consist of polychlorinated biphenyls, pesticides, toxic metals, cyanides, halogenated organics, and non-halogenated volatile organics. The volume of these wastes previously landfilled in California was 500,000 tons/year. The intent of the ban is to force the wastes to be recycled, detoxified, or destroyed, as alternatives to land disposal.

To make this strategy of land disposal restriction more workable for the regulated community, some states such as Illinois specifically require that it must be technically feasible and economically reasonable to require recycling or treatment. The state statutes that require this often authorize the agency to describe in rules and regulations what is technically or economically feasible and what wastes should be restricted from landfills.

Other institutional strategies designed to encourage recovery and reuse alternatives to land disposal include establishment of government operated waste exchanges, research and development programs, and state ownership of alternative technology facilities. Waste exchanges direct material that would otherwise enter the waste stream to a beneficial use or reuse; they lower disposal costs and conserve raw materials and energy necessary to process virgin materials. Research and development can advance the implementation of alternative resource recovery technologies through direct state research or technical assistance programs. Perhaps the most effective institutional strategy is the establishment of state-owned facilities that provide alternative technologies; this permits the regulatory agency to set specific treatment and recycling requirements on the hazardous wastes which will be handled by the facility.

Waste Exchanges

In resource recovery of hazardous wastes it is important to recognize that the generator of a given stream may not necessarily recognize the potential for recovery by an industry outside of its immediate scope of production. The process of secondary recovery can be facilitated by the establishment of waste exchanges.

Waste exchanges are institutions which promote waste reuse by one of two methods. Most waste exchanges act as information clearinghouses, by collecting information on materials to be disposed and potential buyers (or recipients) of such waste. Other waste exchanges act by accepting (or purchasing) wastes from a generator for sale to a user. In the case of a waste materials exchange the acceptance and sale may be merely a paper transaction, as in the case of a brokerage, rather than a physical acceptance and transfer.

The earliest waste exchanges were established in Western Europe, beginning in 1972. These European organizations now typically have a 30-40 percent success rate in exchanging materials contained in their listing. The U.S. waste exchange industry has developed more recently, and its success rate is only about 10 percent.

Constraints on successful waste exchange include long transport distance between the generation and reuse points, and cost of waste purification prior to reuse. Factors which enhance waste exchange include the inherent value of the material, high concentration and purity, quantity and reliability of availability, and high offsetting costs for ultimate disposal.

Due to transportation cost factors, most waste exchanges operate on a local, state or regional basis. Of the top 10 states in terms of volume of hazardous wastes generated, 4 (Illinois, Indiana, Michigan and Ohio) are located in the midwest, as are seven of the 28 known waste exchanges.

Waste exchanges may be operated by government bodies (e.g., Illinois, New York), trade associations such as state chambers of commerce, individual industrial companies to handle their specific wastes, or as private, profit oriented ventures. This latter approach appears most successful, due to both confidentiality aspects and aggressive marketing techniques. Confidentiality is often of industry concern, for two reasons;

1)   to avoid alerting competitors to proprietary information which is perceived to give the generator an economic edge, and

2)   industry may feel that despite good
         intentions, a regulatory body may use
         "inside information" against the
         generator.

Some waste exchanges offer the opportunity for the
generator to approve a proposed user in a waste
exchange activity.

The general categories of wastes which have
been successfully exchanged to date in the U.S. include concentrated acids, alkalis, solvents, catalysts, oils, other combustibles (for fuel value),
and wastes and high concentration of metals.
Other, more specialized materials, have also been
successfully exchanged. Examples include gypsum
wallboard (used as a soil conditioner), scrap roofing shingles and trimmings, and calcium hypochlorite. However, solvents and waste oils appear to
be the most highly sought materials for waste
exchange.

Although on an industry-wide basis it has been
estimated that with available technology only about
3 percent (6 million metric tons/year) of the total
U.S. hazardous wastes volume generated has potential for waste exchange, the percentage in selected
industrial categories is much higher (Table 3.2).
In general, waste exchanges take place from larger
companies using continuous manufacturing processes
to smaller companies using batch processes; from
basic chemical manufacturers to formulator; and
from industries with high purity requirements (e.g.
pharmaceuticals) to those with low purity requirements (e.g. paints).

CONCLUSIONS

Potential recovery strategies for the various
classes of wastes are summarized in Tables 3.3 and
3.4. The areas having the greatest potential for
recovery are:

•   energy recovery from concentrated
         organic liquid wastes, e.g.,
         incineration of waste organic liquids
         and oils

Table 3.2   Categories of Industry Producing
             Hazardous Wastes With Significant
             Potential for Waste Exchange and Reuse
             [USEPA, 1977].

Industry Category	SIC Code	Estimated Percent With Reuse Potential
Pharmaceuticals	2831, 2833	95
Paints and Allied Products	285x	40
Organic Chemicals	2865, 2869	25
Petroleum Refining	2911	10
Small Industrial Machinery	355x	20

- recovery of materials from concentrated organic liquid wastes, e.g., distillation and recovery of waste solvents

- recovery of metals from industrial sludges and metal plating wastes, e.g., recovery of chromium, copper and nickel from spent plating baths

It appears that the following techniques also have potential for development and should be investigated further for possible application for material and energy recovery:

- use of flue gases for pollution control purposes, e.g. neutralization of alkaline wastes

- use of cement kilns for simultaneous destruction of halogenated organic liquids, and energy recovery

- recovery of metal oxides from incinerator ash and other solids

Table 3.3

Summary of Resource & Energy Recovery Potential
for Hazardous Wastes (Inorganic) (Source: Shuster
et al., 1979)

Type of Waste	Direct recovery or reuse	Raw material for secondary use	Energy recovery	Use in pollution control	Low potential for recovery
101-Acid Soln. - no contamin.	x	x		x	
102-Acid Soln. - with metals					
Heavy metals (except Cr)	x	x			
Chromium	x				
Noble metals	x				
Pickle liquor	x	x		x	
103-Acid soln. - with org.					
Emulsified org.		x	x		
Dissolved org.	x	x			
104-Acid sludges					
Inert solids					x
Solids with metals	x				
Organic solids					x
105-Acid gases				x	
111-Alk. soln. - with metals	x	x		x	
112-Alk. soln. - with org.	x	x	x		
113-Cyanide soln.					x
114-Cleaning soln.					x
115-Alkaline sludges					
Inert solids					x
Solids with metals	x				
Organic solids					x
121-Salt solutions					x
122-Solids					x
123-Metals					
Heavy metals	x				
Alkali metals					x
Volatile metals	x	x			
124-Non-metals					
Phos. sulf. compds.	x				
Asbestos					x

Table 3.4

Summary of Resources & Energy Recovery Potential
for Hazardous Wastes (Organic)
(Source: Shuster et al., 1979)

Types of Waste	Direct recovery or reuse	Raw material for secondary use	Energy recovery	Use in pollution control	Low potential for recovery
Concentrated Liquids					
201-Clean, halogenated	x	x	x		
202-Clean, non-halogenated	x	x	x		
203-Clean, solvent mixtures	x	x	x		
204-Dirty, halogenated	x	x	x		
205-Dirty, non-halogenated	x	x	x		
206-Dirty, solvent mixtures	x	x	x		
Dilute Aqueous Solutions					
211-Readily oxidized, halog.					x
212-Readily oxidized, non-halog.					x
213-Difficult to oxid., halog.					x
214-Difficult to oxid., non-halog.					x
Organic Solids					
221-Salts and other solids	x	x			
222-Tars and residues			x		
223-Sludges	x		x		
Organic Gases/Vapors					
231-Combustible			x		
Special Wastes					
311-Strong oxidizing agents					x
312-Explosives					x
313-Biological wastes					x

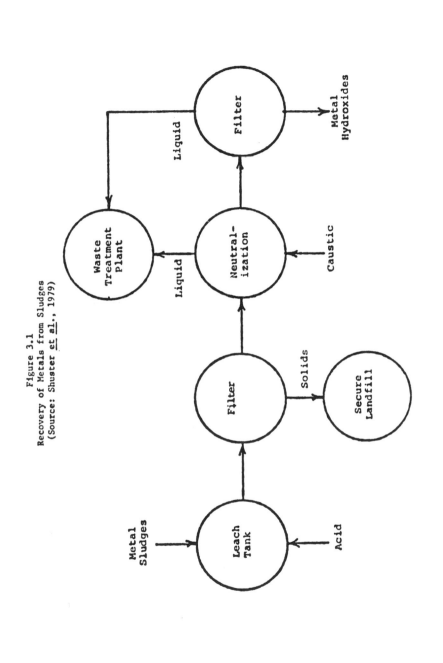

Figure 3.1
Recovery of Metals from Sludges
(Source: Shuster et al., 1979)

Figure 3.2
Recovery of Material & Energy from Waste Solvents
(Source: Shuster et al., 1979)

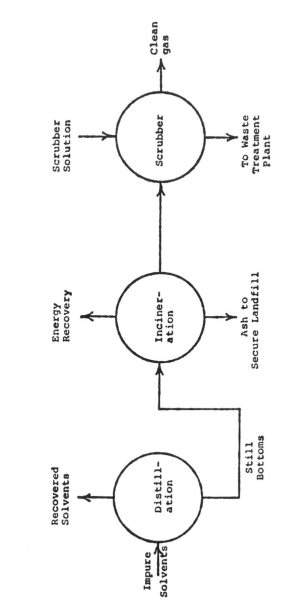

- use of heat exchangers and waste heat boiler for heat recovery

- minimization of auxiliary fuel by mixing wastes of high and low heating value

- use of pyrolysis for fuel recovery and ash fusion prevention

- the development of a waste exchange program

Two typical processing schemes for material and energy recovery are illustrated in Figures 3.1 and 3.2.

Detailed design and analysis of resource recovery facilities are directly dependent on the specific nature, volume and composition of the waste to be treated. Economic feasibility of resource recovery is highly sensitive to such factors as value of recovered products, transportation and storage costs. Transportation and storage at a recovery facility may involve special risks not present at facilities primarily designed for disposal.

REFERENCES

Battelle Memorial Institue, Program for the Management of Hazardous Wastes, Vol. I, EPA/530/SW-54c-1, PB 233 630 (1974).

Besselievre, E.B., "The Treatment of Industrial Wastes", McGraw-Hill Book Company, New York (1969).

Booz Allen Research Inc., "A Study of Hazardous Waste Materials, Hazardous Effects and Disposal Methods", Vol. II, EPA-670/2-73-15, PB 221 466 (1973).

Booz Allen Research Inc., "A Study of Hazardous Waste Materials, Hazardous Effects and Disposal Methods", Vol. III, EPA 670/2-73-16, PB 221 467 (1973).

Boucher, F.B., et al., "Pyrolysis of Industrial Wastes for Oil and Activated Carbon Recovery", EPA 600/2-77-091 PB 270 961.

Cana , J.W., Proceedings of the 14th Industrial
    Waste Conference, Purdue University, p. 26
    (1959).

Federal Environmental Agency (West Berlin),
    "Disposal of Hazardous Wastes, Manual on
    Hazardous Substances in Special Wastes",   NATO
    CCMX Report #55, PB 270 591 (1976).

Fisco, R.A., Proceedings of the 15th Industrial
    Waste Conference, Purdue University, p. 15
    (1960).

Folks, N.E., et al.,   "Pyrolysis as Means of
    Sewage Sludge Disposal", J. Env. Eng. Div.,
    Aug. 1975, EE4 11518 607-621.

Genser, J.M., et al.,   "Alternative for Hazardous
    Waste Management in the Organic Chemical,
    Pesticides and Explosives Industries",
    EPA/530/SW-145c2, PB 276 170, (1977).

Gurnham, C.F.,   "Principles of Industrial Wastes",
    John Wiley & Sons, Inc., New York (1955).

Gurnham, C.F.,   "Industrial Wastewater Control",
    Academic Press, New York (1965).

Hall, E.P., et al.,   "Recovery Techniques in
    Electroplating", Plating and Surface
    Finishing, 65, 2, 49 (1978).

Hitchcock, D.A.,   "Solid Waste Disposal:
    Incineration," Chem. Eng., 185-194, May 21,
    (1979).

Humenick, M.J. and Davis, B.J. "High Rate
    Filtration of Refinery Oily Wastewater
    Emulsions", Jour. Water Poll. Control Fed.,
    50, 1953 (1978).

Iannotti, J.E., et al.,   "An Inventory of
    Industrial Hazardous Waste Generation in New
    York State", New York State Department of
    Environmental Conservation Technical Report
    SW-P14 (1979).

Leonard, R.P., et al.,   "Assessment of Industrial
    Hazardous Waste Practices in the Metal
    Smelting and Refining Industry", Vol. 2,
    EPA/530/SW-145c2, PB 276 170, (1977).

Levin, J., et al., "Assessment of Industrial
    Hazardous Waste Practice – Sepcial Machinery
    Manufacturing Industries", EPA/530/SW-141C, PB
    265 981.

Luthy, R.G., et al., "Removal of Emulsified Oil
    with Organic Coagulants and Dissolved Air
    Flotation," Jour. Water Poll. Control Fed.,
    50, 331 (1978).

Nemerow, N.L., "Industrial Water Pollution –
    Origins, Characteristics and Treatment",
    Addison-Wesley Publishing Company, Reading,
    Massachusetts (1978).

Novak, R.G., "Eliminating of Disposing", Chemical
    Eng. 78, Oct. 5 (1970).

Ottinger, R.S., et al., "Recommended Methods of
    Reduction, Neutralization, Recovery of Disposal
    of Hazardous Wastes", Vol. V, National
    Disposal Site Candidate Waste Stream
    Constituent Profile Reports--Pesticides and
    Cyanide Compounds, EPA-670/2-73-053e, PB 224
    583 (1973).

Ottinger, R.S., et al., "Recommended Methods of
    Reduction, Neutralization, Recovery or
    Disposal of Hazardous Waste", Vol. VI,
    National Disposal Site Waste Stream
    Constituent Profile Reports – Mercury,
    Arsenic, Chromium and Cadmium Components, EPA
    670/2-743-053f, PB 224 585 (1973).

Ottinger, R.S., et al., "Recommended Methods of
    Reduction, Neutralization, Recovery or Dis-
    posal of Municipal Waste", Vol. XIII, Indus-
    trial and Municipal Disposal Candidate Waste
    Stream Constituent Profile Reports--Inorganic
    Compounds,EPA-670/2-73-053m PB 024 592 (1973).

Ottinger, R.S., et al., "Recommended Methods of
    Reduction, Neutralization Recovery or Disposal
    of Hazardous Wastes", Vol. 12.

Patterson, J.W., and Haas, C.N., "Management of
    Hazardous Wastes: An Illinois Perspective",
    Report to the Illinois Institute of Natural
    Resources (1982).

Peters, G.O., et al., "Assessment of Industrial
    Hazardous Waste Practices--Electronic
    Components Manufacturing Industry", US EPA SW-
    140c, NTIS PB 265 532 (1977).

Ross, R.D., "Technology Options in Thermal
    Processing of Organic Hazardous Wastes", AIChE
    Symp. Ser., 73 (162), (1977).

Santoleri, J.J., "Chlorinated Hydrocarbon Waste",
    Chem. Eng. Prog. Vol. 69, No. 1, 68 (1973).

Sebastian, F.P., "Sludge Incineration Solves a
    Sludge Disposal Dilemma", Proc. of the Natl.
    Conf. on Complete Water Reuse, 2nd Waters
    Interface with Energy, Air and Solids, AIChE,
    Chicago, Ill. May (1975).

Shaver, R.G., et al., "Assessment of Industrial
    Hazardous Waste Practices--Inorganic Chemicals
    Industry", EPA/530/SW-104c, PB 244 832 (1975).

Shuster, W.W., et al., "Technology For Managing
    Hazardous Wastes." New York State
    Environmental Facilities Corp., Albany (1979).

Steele, W.R., "Application of Flue Gas to the
    Disposal of Caustic Textile Wastes", Proceed-
    ings of 3rd Southern Municipal and Industrial
    Waste Conference (1954).

Steiner, J.L., et al., "Pollution Control
    Practices: Air Floatation Treatment of Refinery
    Waste Water", Chem. Eng. Progr., 74, 12, 39
    (1978).

Tabakin, R.B., et al., "Oil/Water Separation
    Technology: The Options Available", Water and
    Sewage Works, 125, 7, 74, and 72 (1978).

Terry, R.C., et al., "Waste Clearinghouses and
    Exchanges: New Ways for Identifying and
    Transferring Reusable Industrial Process
    Wastes", Arthur D. Little, Inc., US
    Environmental Protection Agency, PB 261 287
    (1976).

"Report to Congress: Disposal of Hazardous Wastes",
    US EPA SW-115 (1974).

"Process Design Manual for Suspended Solids
     Removal" U.S. Environmental Protection Agency
     - Technology Transfer.   EPA 625/1-75-003a
     (1975).

Wadhawan, S.C.,   "Economics of Acid Regeneration -
     Present and Future", Iron and Steel Eng., 55,
     10 48 (1978).

Weber, W.J. Jr.,   "Physicochemical Processes for
     Water Quality Control", Wiley-Interscience,
     New York (1972).

CHAPTER 4

DEVELOPING TECHNOLOGY FOR
RECOVERY, RECYCLE, OR REUSE

INTRODUCTION

In addition to the established technologies
discussed elsewhere in this book for recovery,
there are a number of developing technologies with
significant potential. Among these are supercritical fluid extraction, electrolytic demulsification,
and extraction via liquid surfactant membranes.
This chapter describes these emerging technologies,
and provides examples of possible future application.

DEVELOPING TECHNOLOGIES

Supercritical Fluid Extraction

Supercritical fluid extraction is a new separation technique which has received much attention
[Chem. Eng., 1979]. Although conditions vary for a
given application, the supercritical fluid extraction process has two fundamental operations. In
the first step, the fluid, usually at a pressure of
4,000-5,000 psi, and a density comparable to liquid
hexane, flows through the material to be treated in
an extraction vessel. At this elevated pressure
its solvation power is greatly enhanced, and the
fluid becomes solute-laden. The fluid is then de-
pressurized in a separation tank. Because the
solvation power of a supercritically compressed
fluid varies with respect to pressure, some of the
extract will drop out of solution. Little solvent
residue is generally left behind in the extract.

According to one manufacturer, the process can
extract materials such as oils from natural products, organic pollutants from wastewater, aromatic
isomers from mixtures, low-molecular weight materials from polymers, and light components from coal

[Chem. Eng., 1982]. It has also been applied to
the regeneration of wastewater adsorption beds
[Eckert, undated]. Among the solvents which can be
used in the system are carbon dioxide, water, oxygen, ethlene, propane, and propylene [Chem. Eng.,
1982].

Actual separation pressures, which can range
from 500 to 3,000 psi, depend on the economics of
the application. Because a greater pressure change
requires more energy than a small one, a tradeoff
exists between the benefit of total solute recovery
by depressurization to ambient conditions and the
cost of repressurizing the solvent for another pass
through the system [Chem. Eng., 1982].

Electrolytic Demulsification

In the future, physical separation techniques
will be required which have capabilities surpassing
those of conventional technology for separating
intimate liquid-liquid and liquid-gas mixture [EPA,
1979]. In several large-scale industrial processes, the liquid-liquid mixtures occur in the form
of micro-emulsions of organic liquids in aqueous
solutions. The treatment of these emulsions by
conventional means would require large batch operations and generate voluminous quantities of sludge
from which recovery is uneconomical. There have
recently been some promising developments in the
area of emulsion-breaking.

It has been shown that passing the micro-
emulsion through a packed-bed anode of iron (or
low-carbon steel) chips in which ferrous ions are
elctrochemically generated is an extremely effective destabilization technique. The process requires simultaneous oxygenation by sparging or by
electrolytic evolution from a carbon anode
[Weintraub and Gealer, 1977; Weintraub,
Dzriechiuch, and Gealer, 1976]. Such a process
would have application in the treatment of oily
wastewater streams generated by the use of cutting
oils in industrial matching operations [Gealer,
Golavou, and Weintraub, undated]. Other possible
applications include tertiary oil recovery, where
alkali and surfactants are injected into wells and
metal recovery from low-grade mineral resources
[Tavlarides, 1981], where solvent extraction is
used as a concentration device or as a pollution
control step.

The mechanism by which the destabilization of micro-emulsions occurs in the electrolyte process is only partially understood (Dzieciuch, Weintraub and Gealer, 1976], although the incentive for more research into the development of a continuous process for the treatment of oil-in-water emulsions exists. The process should have as a primary requirement not only that it is effective in destabilizing micro-emulsions but also that the sludge it generates should be as "dry" as possible to facilitate recovery of organics.

Liquid Surfactant Membranes

Liquid membranes are formed when an emulsion of two immiscible phases is dispersed in a third (continuous) phase. The encapsulated phase of the emulsion is thus set apart from the continuous phase by the second phase of the emulsion, which is immiscible with either the encapsulated or continuous phases and constitutes, within this context, the liquid membrane. The stability of the emulsion is maintained by the addition of surfactants and stabilizing agents.

When developed for a given application, the three-phase system may be used to facilitate separation of components in a mixture by selective diffusion of the desired component from the continuous phase to the encapsulated phase, or vice versa. Following this transfer, the three phases can be separated by first settling and separating the emulsion and the continuous phase, and then breaking the emulsion [Wasan, 1982].

Liquid surfactant membranes have been shown to be of use in recovering phenol, ammonia, organic acids, amines, and hydrogen sulfide from wastewater [Cahn and Li, 1974; Cahn, Li, and Miday, 1978; Li, Cahn, and Shrier, 1973]. The separation of inorganic ions such as chromium and copper from solution has also been studied [Hochhanser, 1975; Strzelbicki and Charewica; Senkan and Stauffer, 1981]. Use of this technology is curbed by a lack of information pertaining to mechanisms which can be used to maximize flux into the capacity within the receiving phase, and an understanding of the relative impact of process variables on these mechanisms [Wasan, 1982].

Dissolution

Dissolution may be defined as the complete or partial transfer of one or more components from a solid phase into a liquid phase in contact with the solid. The reaction involves some degree of chemical transformation (solvation, ionization, oxidation) of the species being dissolved [Noyes Data Corp., 1978]. Although dissolution reactions are used in nearly all areas of chemical processing, there has been recent interest in the in-situ dissolution of uranium from ore [Tavlarides, 1983] and of bitumen from tar sands [Shahinpoor, 1982].

In the case of in-site leaching of uranium from underground deposits, the subsurface deposit is flooded with leach solution and then pumped to the surface ready for uranium recovery. Advantages of this method are a significant reduction of processing costs and minimal disturbance to the surface conditions, as well as production of a relatively small volume of waste requiring disposal [Tavlarides, 1983].

The goal of in-situ bitumen recovery is to reduce the viscosity of the bitumen while it is in place, by heating and/or diluting it, and to subsequently collect the liquefied product. Several solvents capable of reducing oil viscosity are available, although they are generally more expensive than the oil produced. Economic success, therefore, requires a high percentage of solvent recovery, which is often difficult to achieve. In certain cases, water may be a suitable solvent [Shahinpoor, 1982].

The in-situ dissolution of uranium is feasible only where the ore body is contained within a rock formation which is relatively impermeable. Groundwater contamination may otherwise be a problem [Tavlarides, 1983]. Potential environmental impacts of producing oil from tar sands appear, on the basis of methods tested to date, to be similar to those of conventional oil field operations [Shahinpoor, 1982].

Particle Conditioning

In the presence of water, solids have an intrinsic electrical surface charge. Almost all matter dispersed in spent process water such as oil particles, salt, biocolloids, inorganic colloids, etc., has a negative charge which is repelled by the negative electrical surface charge of granular media. In order to maximize filtration efficiency, these coulombic repulsive forces must be regulated through control of the physicochemical properties of the dispersed solids. The colloids must be destabilized into agglomerates tough enough to resis redispersive hydraulic forces in the filter.

Historically, granular media filtration has been viewed as a polishing step following a clarifier. More recently, direct filtration of highly contaminated wastewaters has been investigated. Results indicate a large savings in capital, chemical, and sludge treatment costs may be realized. Charge neutralization or reversal by adsorption of a destabilizing chemical to the colloid is a key mechanism for optimization of direct filtration. When molecules of the added chemical attach to two or more colloids, aggregation and bridging occurs, and agglomerates are formed which resist redispersion.

The processes are demonstrated in the case of removal of coke fines from hydraulic decoking water. Coke fines are originally stabilized by negative zeta potential. This charge is easily reversed by the addition of cationic polyelectrolytes. The polyelectrolytes, which also have good bridging properties, cause most of the solids to be enmeshed in a polymer. Upon addition of a small amount of weakly anionic polyelectrolyte, the positively charged particles are "collected" into massive, easily separable aggregates. Once the aggregates are formed, the forces which bind them are strong enough to resist redispersion by the hydraulic forces of direct filtration [Grutsch and Mallott, 1977].

REFERENCES

Cahn, R.P., Li, N.N. and Miday, R.M., "Removal of Ammonium Sulfide from Waste Water by Liquid Membrane Process," Env. Sci. and Tech., Vol. 12, p. 1051, 1978.

Cahn, R.P. and Li, N.N., "Separation of Phenol from Waste Water by the Liquid Membrane Technique," Separation Sci. 9(6), p. 508-518, 1974.

Chem. Eng., Mar. 12, 1979, pp. 41-43.

Dzieciuch, M.A., Weintraub, M.H., and Gealer, R.L., Ext. Abst. No. 260, 149th Mtg. Electrochemical Society, Washington, D.C., May 2-7, 1976.

"Electrolytic Treatment of Oil Waste Water from Manufacturing and Machining Plants," by R.L. Gealer, A. Golavou, and M. Weintraub (Ford Motor Co.) Report on Grant No S804174, IERL/EPA (Cincinnnati, OH).

"Environmental Considerations of Selected Energy-Conserving Manufacturing Process Options" (20 Volumes), A.D. Little, Inc., EPA-60017-76-034, Dec. 1976 - Aug. 1979.

Grutsch, J.F. and Mallott, R.C., "Optimizing Granular Media Filtration" CEP., April 1977, pp. 57-66.

Hochhanser, A.M. "Concentrating Chromium with Liquid Surfactant Membranes," AIChE Symp. Ser., 71(152) p. 136-142, 1975.

Identification of Research and Planning Needs for Industrial Waste Management. IWERC Report, Mar. 16, 1981.

Li, N.N., Cahn, R.P. and Shrier, A.L., U.S. Patent 3,779,907, Dec. 18, 1973.

Noyes Data Corp. Unit Operations for Treatment of Hazardous Industrial Wastes. Park Ridge, N.J. 1978.

Patterson, J.W. and Haas, C.N., Management of Hazardous Wastes: An Illinois Perspective (Draft), report to the IllinoisInstitute of Natural Resources, 1982.

Pojasek, R.B., "Solid-Waste Disposal: Solidification." Chem. Eng. August 13, 14 (1979).

Progress Report on Research Planning Task Group Studies-Separation Technology. C.A. Eckert to R.S. Engelbrecht.

Senkan, S.M. and Stauffer, N.W., "What to do with Hazardous Waste." Technology Review 84(2), pp. 34-47, 1981.

Shahinpoor, M. "Making Oil from Sand" Technology Review 85(2):48-54, 1982.

Strzelbicki, J. and Charewica, W. "Separation of Copper by Liquid Surfactant Membranes, J. Inorg. Nucl. Chem. Vol. 40, p. 415-421.

"System is Designed for Critical-Fluid Extractions," Chem. Eng., January 25, 1982, pp. 53.

Tavlarides, L.L., "Process Modification Towards Minimization of Environmental Pollutants in the Chemical Process Industry" Final report submitted to IWERC, 1983.

Tavlarides, L.L., "Refining of Non-Ferrous Metals" and "Electroplating," Bi-Monthly Report IWERC, Mar. 16, 1981.

Wasan, D.T., "Separation of Metal Ions by Accelerated Transport Through Liquid Surfactant Membranes." Preproposal submitted to IWERC. January 8, 1982.

Weintraub, M.H., Dzrieciuch, M.A., and Gealer, R.L., Ext. Abstr. No. 261, 149th Mtg. Electrochemical Society, Washington, DC., May 2-7, 1976.

Weintraub, M.H., and Gealer, R.L. "Development of Electrolytic Treatment of Oily Wastewater," AIChE 70th Annual Mtg. N.Y., Nov. 13-17, 1977. Paper No. 151.

INTRODUCTION

The term sorption includes both adsorption and absorption and refers to a process in which a solute (mixture component) moves from one phase and is accumulated in or onto another. Adsorption occurs when material is collected at the interface between two phases. Adsorption can occur between a liquid-liquid, gas-liquid or liquid-solid interface [Weber, 1972]. In absorption, the two phases are brought into contact so that the mixture components can diffuse from one phase to another. During the contact of the two phases, the components of the original mixture redistribute themselves between the phases.

According to Geankoplis (1978), there are many chemical process materials and biological substances that occur as mixtures of different components in the gas, liquid or solid phase. If it is desired to remove one or more of these components from its original phase, another phase must be contacted.

ABSORPTION

The absorption of gases in liquids is one of the most frequently used techniques for controlling the composition of industrial waste gases prior to their discharge to the atmosphere. Waste gases are generally a mixture of gaseous components, some of which are soluble in a selected liquid phase. Direct contact of gas with liquid enables mass transfer to take place between the phases, in directions governed basically by the concentration gradients of the individual components.

The rates of mass transfer between the gas and adsorbent are determined mainly by the amount of

surface area available for absorption. Other factors governing the rate of absorption, such as the
solubility of the gas in the absorbent and the
degree of chemical reaction, are characteristic of
the constituents involved and independent of the
equipment used [US EPA, 1978].

Mass transfer between two fluids is carried
out by eddy diffusion in the bulk of each phase and
by molecular diffusion close to the interphase
boundary. It is assumed that complete equilibrium
is established instantaneously at the boundary, and
that from this boundary, active species are transported deeper into each phase by molecular diffusion followed at some depth by mixing of isolated
elements caused by eddy currents in the bulk phase.
Eddy diffusion is orders of magnitude faster than
molecular diffusion. The overall mass transfer
rate is therefore controlled by mass fluxes within
the molecular diffusion layers.

During the contact period, active species
diffuse into an element, developing a concentration
gradient which changes both with time and position.
[Bettelheim, et al., 1978]. In most cases, absorption of one component is simultaneously accompanied by the desorption of another. Deliberate
desorption can be achieved by a change of physical
conditions, stripping of the liquid with inert gas
(or steam), or by chemical decomposition of the
sorbent. Absorption followed by desorption constitutes a cyclic operation which allows reuse of the
sorbent and acts as a device for separation and
concentration of the selected gas. [Bettelheim, et
al., 1978].

Gas absorption equipment is designed to provide thorough contact between the gas and liquid
solvent, to permit interphase diffusion of the
materials. Contact may be provided by various
types of equipment, the most common being plate and
packed towers.

Plate towers employ stepwise contact. Several
plates or trays are arranged so that gas is dispersed through a layer of liquid on each plate.
The number of plates required is dictated by the
difficulty of the mass transfer operation and the
desired degree of absorption.

Packed towers are filled with a packing material having a large surface to volume ratio. The packing is wetted by the absorbent, providing a large liquid film surface area for continuous contact with the gas.

The flows through plate and packed towers are usually countercurrent, with the liquid being introduced at the top and the gas at the bottom. This arrangement results in the highest possible transfer efficiency.

Spray towers and venturi scrubbers have more limited application to gas absorption. Spray towers dispense liquid absorbent in a spray through which the gas is sent, while venturi scrubbers contact the gas and the absorbent in the throat of a venturi nozzle. The gas-liquid mixture then enters an entrainment separator in which centrifugal force separates the liquid droplets from the gas.

Packed and spray towers introduce lower pressure losses than do plate towers, and there is a high pressure drop associated with the operating velocities of the venturi scrubber. Power requirements for venturi scrubber operation are consequently large. Although spray towers have the advantage of removing particulate matter without plugging, they provide the least effective mass transfer capability and their use is restricted to applications requiring only limited removal of highly soluble gases. Venturi scrubbers, also highly efficient for particulate removal, are preferred for removal of a highly soluble gas.

Because spray towers and venturi scrubbers have limited application to absorption, packed or plate towers are usually the equipment of choice. A comparison of these two types of equipment is given in Table 5.1 [U.S. EPA, 1978].

Limiting Technology

In general, absorption is most efficient under the following conditions:

1.   The vapors to be absorbed are quite soluble in the absorbent.

2.   The absorbent is relatively nonvolatile.

3.   The absorbent is noncorrosive.

4.   The absorbent has low viscosity.

5.   The solvent is nontoxic, nonflammable, chemically stable, and has a low freezing point.

Table 5.1   Comparison of Packed and Plate Towers [US EPA, 1978]

---

1.   Packed towers are less expensive than plate towers when materials of construction must be corrosion resistant.

2.   Packed towers have smaller pressure drops than plate towers designed for the same throughput.

3.   Packed towers are preferred for foamy liquids.

4.   Packed towers usually have a smaller liquid holdup than plate towers.

5.   Plate towers are preferred when the liquid contains suspended solids since plate towers are more easily cleaned.

6.   Plate towers are preferred for larger installations because they minimize channeling and reduce tower height.

7.   Plate towers are more suitable when the process involves appreciable temperature variation, since expansions and contractions due to temperature changes may crush the tower packing.

8.   Plate towers are preferred when heat must be removed, because cooling coils are more easily installed.

9.   Packed towers are preferred in sizes up to 2 feet in diameter if other conditions are nearly equal.

The rate of mass transfer between the absorbent and the gas is dependent on the amount of surface area available for absorption. The solubility of the gas in the absorbent and the degree of chemical

reaction which take place are characteristic of the
constituents involved. Gas absorption equipment
must be designed to provide adequate contact between the gas and absorbent liquid to permit
interphse diffusion of the organic vapors [US EPA,
May 1978].

     Selection of a suitable liquid solvent and the
determination of the limits of absorption efficiency are based on solubility data. In most cases,
formulae are not available for calculation of the
solubility and only tabulated or graphical data can
be used.

Recycle, Recovery, Reuse Applications: Organics

     Many industrial processes which employ phenol
or phenolic resins are faced with the problem of
toxicity and odor problems arising from the evolution of phenol and phenol-derived volatile substances into the air. A method developed for substantially reducing this atmospheric contamination
consists of absorbing the contaminants in an
aqueous solution of a water soluble phenol-
formaldehyde resin, and then condensing the mixture
of absorbing resin and absorbed contaminants to
form an augmented resin by-product [Baker, 1975].

     Gases contaminated with vapors from volatile
organic liquids may be recovered by contacting the
vapor-containing gas with an absorbent sponge oil
in an absorbing tower. The sponge oil, rich in
absorbed vapors, is conveyed to a flash tank where-
in the absorbed vapors are removed and recovered.
Following this process, the sponge oil can be successfully reused [Haines, 1975].

     Nichols (1973) describes a low-temperature
recirculating absorption system capable of 90
percent hydrocarbon recovery and designed to recover vapors present in saturated vapor-air mixtures
from the loading and storing installations of oil
companies. The system consists of an absorber and
components which condition the vapors and liquid,
improve absorption efficiency, reduce thermal losses, and/or improve system safety. In the system,
incoming vent gases are saturated with fuel and the
entrained liquid is allowed to settle out before
passing to an air compressor. The gases are
brought to 45 psig and approximately $300^{\circ}F$ as they
pass from saturator to compressor. They then

proceed to an after cooler where temperature is reduced to an ambient level. From the aftercooler,
they pass through a bubble bar and into an absorber
where they are absorbed by sprayed gasoline.

Recycle, Recovery, Reuse Applications: Inorganics

When plastic wastes are burned in an incinerator, they tend to exhaust a dark smoke resulting
from incomplete combustion. In the case of polyvinyl chloride (PVC) incineration, serious corrosion of equipment as well as air pollution by
hydrogen chloride gas will occur. Oda, et al.
(1975) devised a disposal system which results in
recovery of hydrochloric acid from PVC wastes. The
process includes carbonization of plastic wastes
for about 40 minutes at 280°C to 300°C to remove
the chlorine from vinyl chloride. Hydrochloric
acid generated by this step is absorbed by water.

CO_2 and H_2S may be removed using aqueous
ammonia, with production of ammonium sulfate which
may be recovered for sale as a fertilizer. Mono-
ethanolamine is used to remove CO_2 from flue gas
for its recovery or for the purification of flue
gas.

Within the past several years, a great deal of
work has been done with gas/solid absorption
systems, particularly with regard to sulfur dioxide
(SO_2) removal. These processes should not be considered for streams of low flow, however. SO_2 can
be removed by liquid absorption and purified for
later collection, compression, and resale. The
Sulphidine process uses a mixture of tylidine and
water, and produces by-products of sodium sulfate
and pure SO_2. In another process designed to
remove SO_2 from smelter gas, dimethylaniline is the
absorbent used to remove the gas from the stream,
producing a by-product of either dilute sulfuric
acid or liquid SO_2 [Ross, 1972].

In the alkalized alumina process developed by
the U.S. Bureau of Mines, flue gas is passed
through a fluidized bed of alkalized alumina. The
bed reacts with the SO_2 in flue gas at temperatures
of 300 to 350°C, resulting in its conversion to
sulfate. After treatment with a suitable reducing
gas, hydrogen sulfide (H_2S) gas is produced. The
H_2S is processed through a Claus plant to bring
about its conversion to elemental sulfur [Ross,

1972].

ADSORPTION FOR AQUEOUS SYSTEMS

Adsorption on activated carbon has been known for a long time [Hasler, 1974], however, its application in the field of municipal and industrial wastewater treatment has become common only in recent years.

The term "activated carbon" applies to any amorphous form of carbon that has been specially treated to give high adsorption capacities. Physically, there are two forms of activated carbon: powdered and granular. The former include particles that are less than U.S. Sieve Series No. 50, while the latter are larger [US EPA, 1971a]. Typical raw materials from which activated carbon is made include coal, wood, coconut shells, pulp mill residues, petroleum base residues, and char from sewage sludge pyrolysis. These carbon materials are activated through a series of processes which include [Cheremisinoff and Morressi, 1978]:

1. Removal of all water (dehydration);

2. Conversion of the organic matter to elemental carbon by driving off the non-carbon portion (carbonization); and

3. Burning off tars, methanol, and other by-products at high ($750-950°C$) temperatures (activation).

Adsorption involves the interphase accumulation or concentration of substances at a surface or interface [Weber, 1972]. The process can occur at an interface between any two phases; such as, liquid-liquid, gas-liquid, gas-solid, or liquid-solid. The material being adsorbed is called the adsorbate while the adsorbing phase is termed the adsorbent.

There are two types of adsorption: physisorption and chemisorption. The former occurs because of weak Van der Waal's forces while chemisorption is a result of a chemical interaction between the adsorbent and adsorbate. Physisorption is characterized by a relatively low energy of adsorption and may be multilayered. Chemisorption processes, however, exhibit high energies of adsorption.

The adsorption process can be affected by many factors. Some of these include [Cheremisinoff and Morressi, 1978]:

1. The physical and chemical characteristics of the adsorbent, i.e., surface area, pore size, chemical composition, etc.

2. The physical and chemical characteristics of the adsorbate, i.e., molecular size, molecular polarity, chemical composition, etc.

3. The concentration of the adsorbate in the liquid phase.

4. The characteristics of the liquid phase, i.e., pH, temperature, etc; and

5. Residence time of the system.

Factors Affecting Adsorption: Inorganics

The surface area of the activated carbon plays a critical role in the removal of inorganics by adsorption. It may typically range from 500 to 1400 m^2/g, with some carbons having surface areas up to 5,500 m^2/g. Another important parameter affecting the adsorption process is the chemical nature of the surface of carbon. This chemical nature varies with the carbon type, and can influence attractive forces between molecules. For the most part, activated carbon surfaces are nonpolar making the adsorption of inorganics difficult [Cheremisinoff and Morressi, 1978].

Regeneration Operation: Inorganics

Wastewater treatment with activated carbon involves two major separate process operations, namely, contacting and regeneration. During the first operation, the water is contacted with the carbon. Impurities are removed from the water by adsorption to carbon. After a period of time, the adsorptive capacity of the carbon is exhausted. Then the carbon is taken out of service and regenerated, usually, by combustion. During this process, the impurities are adsorbed for potential recovery.

On some systems an additional process operation, backwashing, may be necessary. It is
required in cases where suspended solids are trapped in carbon beds causing severe head losses.
Backwashings are more frequent for downflow contactors, which may be designed for suspended solids
removal also.

Regeneration of spent carbon is practiced in
cases where a large quantity of carbon is used.
Thermal reactivation is the most common method used
for regeneration of carbon. Other methods include
alkaline regeneration for acid adsorbates, acid
regeneration for basic adsorbates, stream regeneration and solvent regeneration.

Recycle, Recovery and Reuse Applications:
Inorganics

The major applications of activated carbon in
wastewater treatment for inorganics include:

1.   Silver and Gold Processing Plants
      [Cheremisinoff and Morrissi, 1978].

2.   Inorganic Chemical Industry [Suisi, et
      al., 1970; Taushkano, et al., 1974].

3.   Electroplating Industry [U.S. EPA,
      1971b; Smithson, 1971].

4.   Refineries [Loop, 1975].

5.   Various Industries With Metal-laden
      Effluents [Linstedt, et al., 1971;
      Cheremisinoff and Habib, 1972; Netzer,
      et al., 1974].

Various inorganic substances from the above-
listed industries are removed from the effluents
and possibly recovered through activated carbon
treatment. The degree of adsorption of these inorganics on carbon and the final recovery vary from
species to species. The major group of inorganics
which are recovered through activated carbon treatment are metals. By-product recovery is advantageous for valuable substances and also in cases
where regeneration of carbon results in low recovered adsorptive capacity of carbon.

Factor Affecting Adsorption: Organics

The adsorption capacity solutes is thought to be a function of:

1.   Adsorbate properties such as functionality, branching or geometry, polarity, hydrophilicity, dipole moment, molecular weight and size, and aqueous solubility.

2.   Solution conditions, including pH, temperature, pressure, adsorbate concentration, ionic strength, and the presence of background and competitive solutes.

3.   The nature of the adsorbent, such as surface area, pore size and distribution, surface distribution, and surface characteristics [Miller, 1980a].

Adsorbents-Activated Carbon and Resins: Organics

Little quantitative information is known about the surface characteristics of carbon and its influence on organic adsorption selectivity. These characteristics are important to adsorption of specific solutes, however. Two types of surface interactions are thought to predominate. The first is that of van der Waal force interactions, hydrophobic in nature, and occurring on a majority of the surface. The second type occurs at the more reactive edges, and may be characterized by positive physical and perhaps chemical interactions due to hydrogen bonding and electrostatic forces. This second type of interaction occurs at a small fraction of the total surface area. Specific adsorptions will result from the presence of oxides, hydroxyls, and other groups on the surface. Activated carbons produced by different processes probably differ in their adsorptivity as a result of their different energy potential and the extent of their heterogeneous sites [Miller, 1980a].

Snoeyink, et al. (1979) found that the nature of sorption sites can vary significantly between different carbons, even though they have similar total surface areas. Their results indicate that phenol sorbs more extensively on a coal-based carbon than on a coconut-shell carbon, at low

phenol concentrations. Recommendations were made
for study in the areas of the sorptive behavior and
characteristics of activated carbon, the types of
functional groups on the surface, and possible
alteration of the surface to produce a more efficient adsorber for a given purpose.

El-Dib, et al. (1979) states that little is
known about the adsorption characteristics of soluble aromatic hydrocarbons on granular carbon and
its efficiency in the removal of such organics. In
a study of benzene, toluene, o-xylene and ethylbenzene, they found that these compounds were adsorbed
in accordance with the Freundlich model, and that
the Freundlich parameters K and n reflect the effects of chemical structure, solubility, and competitive interactions on the adsorption process. In
the case of a mixed-solute solution, they found
that uptake of each compound was considerably reduced, although the order of adsorption was the
same.

The use of polymeric adsorbent resins should
be considered in cases where the economics of solvent or chemical regeneration of the adsorbent is
favorable. Polymeric adsorbent resins are similar
in size, shape, and appearance to conventional ion
exchange resins, but differ in the respect that
they contain no ionically functional sites. There
are two basic families of adsorbent polymers available: one is based on a crosslinked polymethacry-
late structure, while the other is based on a
crosslinked polystyrene structure [Fox, 1979].
Although capital costs of synthetic adsorbent
systems and those of activated carbon are comparable, operating costs indicate that polymeric
adsorbent methods are more economical than carbon
systems when the level of dissolved organics is
high [Stevins and Kerner, 1975]. An attractive
feature of synthetic resins is that they can, at
least theoretically, be designed and manufactured
for a specific adsorption application [Kim, et al.,
1976].

It has been demonstrated that Amberlite XAD
polymeric adsorbents can remove substantial quantities of phenolic compounds from aqueous solutions,
with the added benefit of being easily regenerable
with nonaqueous solvents or caustic solution. This
permits the recovery of a useable form of phenolic
material in many cases. Crook, et al. (1975)

studied the results of flow rate, concentration,
temperature, and pH on phenol removal. They found
that at 6700 mg/l influent phenol concentration,
the capacity of Amberlite XAD-4 polymeric sorbent
for phenol is (to 1 mg/l leakage) 87 g/l, while at
300 mg/l influent concentration the capacity decreases to 72 g/l. Cumulative phenol leakage was
found to be slightly higher at 5°C than at 25°C.
Although flow rates were varied from rates of 2 to
4 bed volumes per hour, the resultant phenol leakage was 0.1 mg/l or less up to the break point. pH
values of 3 and 6.45 were used to determine the
effect of pH on phenol removal. Results indicated
that slightly better performance may be obtained in
the lower pH range. In another experiment Crook,
et al. (1975) tested the effect of bed depth on
removal of p-nitrophenol and found that a 15 inch
bed depth column of XAD-7 resin does not afford
substantial improvement over a 9 inch bed depth in
either leakage or capacity at the flow rate of
effluent studied. Removal of Bisphenol A was also
studied using both the XAD-4 and XAD-7 resins.
These resins differ in polarity. It was found that
the XAD-4 resins successfully treated 33.5 bed
volumes while the XAD-7 treated 16 volumes. These
experiments are mentioned to illustrate the many
variables which may or may not effect removal of an
organic compound from an aqueous solution.

     The use of carbon adsorption and, more recent-
ly, polymeric resin adsorption for removal of organic contaminants from wastewater have proven to
be effective. However, DiGiano (1980) states, with
reference to carbon adsorption, that there are many
unanswered questions regarding control of process
performance and that part of the problem originates
from the variability in composition and concentration of contaminants in waste streams. Regarding
polymeric resins, complex reactions involved in the
adsorption process can only be partially predicted.
Laboratory feasibility studies and, in some cases,
pilot studies under actual use conditions, are
generally needed before the appropriate resin and
operating conditions for a specific application are
defined [Fox, 1978].

Regeneration and By-Product Recovery: Organics

     Himmelstein, et al. (1973) reviewed the various
methods of in-place regeneration of activated
carbon. Reactive regeneration, by which phenol is

desorbed from the column by reaction with caustic
soda, has been practiced. In the plant where this
took place, the regenerant solution was suitable
for recycle without further treatment and an excel-
lent example of an opportunity for recovery.
Phenol is recovered in the product stream, and the
residual caustic is used as a component in phenol
production.

Solvent regeneration involves the use of a
solvent phase to desorb the organics from the carbon. The solvent may be recovered for reuse in
subsequent regeneration while the desorbed materials may be reintroduced into the process or re-
fined for reuse or sale. In some cases, the regenerated steam may be recycled without further processing. Unlike reactive regeneration, solvent
regeneration is feasible in cases where direct
recycle of a regenerant stream is not feasible.
Separation of the solvent and recovered material
may be accomplished by distillation, extraction,
decantation, or precipitation. Recovery by steam
treatment has been demonstrated by laboratory
studies to be feasible for waste streams such as
those containing acetic acid, aromatic acids, chlorinated aromatics, phenols, alcohols and esters.

Due to the high binding energies of carbon,
thermal reactivation of activated carbons is often
the only successful means of regenerating the carbon. Thermal regeneration also normally leads to
destruction of the adsorbed species, making their
recovery impossible. Baker, et al. (1973), however, describes the use of carbon adsorption of
sulfite effluent prior to oxidation, followed by
chemical regeneration of cresylics and return of a
valuable product to the process in a full-scale
plant situation.

Jensen (1980) describes a semi-continuous
activated carbon system for removal of dissolved
acrylic monomers from wastewater. By using such a
unit, acrylic monomers can be recovered at some
value, and the wastewater is upgraded for reuse.
Three carbons were compared for suitability for
these wastewaters.

Phenolic compounds are bound to polymeric
resins by van der Waal's forces. Parmele (1979)
has shown this by measuring the enthalpy of binding
mono and dichlorohenols to Amberlite XAD-4 (-4 to

-6 kcal/mole). The enthalpy range is within
the range for physical forces holding the phenolic
to the adsorbent. Small energy inputs are required
for desorption, and the phenols can thus be easily
recovered by regenerating the resin. This is the
key to the value of resins for waste material
recovery. Fox and Himmelstein (1974) discuss several applications of polymer adsorption and regeneration for the recovery and recycle of phenol,
para-nitrophenol and phenoxy acid pesticide.

Stevens and Kerner (1975) state that the binding energies of synthetic resins are lower than
those of activated carbon for the same organic
molecules. This would make recovery of organics
from resins more attractive than from carbon, due
to smaller energy inputs needed for desorption.
Fox (1979) lists organic solvents including ace-
tone, methanol, isopropanol, and inorganic solvent
systems such as steam, aqueous caustic solutions,
and aqueous acids for resin regeneration.

Chlorinated pesticides such as endrin, DDT,
2,4-D, toxaphene, and polychlorinated biphenyls can
be efficiently removed by adsorption onto an adsorbent resin with polystyrene structure. One plant
which manufactures 2,4-D, and related herbicides
practices recovery of phenoxy herbidicides and
their intermediates for recycle to the process.
Regeneration is performed with saturated steam.
The resin is regenerated with methanol, and the
herbicide and its intermediate are recovered by
distillation.

Aqueous effluents from vinyl chloride and
other chlorinated hydrocarbon manufacturing plants
contain up to 1 percent of a mixture of ethylene
dichloride, chloroform, and/or carbon tetrachloride. After steam regeneration of the resin, the
collected condensed organic phase can be either
reused in the plant or incinerated. Effluent
streams from the production of benzene, toluene,
and xylene are usually contaminated by these organics. Recovery of these compounds is carried out
in a process scheme similar to that for chlorinated
hydrocarbons. From these examples, it is evident
that polymeric adsorbents may be used by industry
for removal and recovery of chlorinated pesticides,
phenols, aliphatic chlorinated hydrocarbons, benzene, toluene, and xylene [Fox, 1979].

Predicting Adsorption: Organics

Several approaches have recently been developed to relate characteristics of molecular structure or a chemical property such as solubility to adsorption potential. Manes and Hofer applied the Polanyi adsorption potential theory, originally developed to describe gas-solid adsorption. Follow up studies have shown this technique to be generally successful for describing single, bi- and tri-solute equilibria, but testing is still limited. Problems with the application of the theory include the fact that the solubilities of many organic compounds of interest are not readily available, while other compounds are relatively soluble, requiring that a more complicated model be used [DiGiano, 1980].

Although few multicomponent studies have been done, equilibrium models have been developed. These include the Langmuir Competitive Model, which maintains that adsorption sites are available to all adsorbates; the Langmuir Semi-Competitive Model, which states that some sites are available to all absorbates while others are only available to certain adsorbable material; and the Ideal Solution Theory, which is developed from thermodynamic considerations. Each model was tested on its own bisolute systems, and little has been done to compare the models on additional systems. In a column study, all three were unsuccessful in predicting relative loadings.

It may be possible to simplify the description of competitive systems by grouping adsorbable components into a few broad classes according to their equilibrium adsorption behavior. A single, synthesized equilibrium isotherm for each class would be used in simulation of competitive adsorption. For this to occur, a method must be found to account for competitive behavior which does not conform to that expected based upon single component isotherm. This occurred in a phenol-dodecylbenzenesulfonate (DBS) system. Phenol was absorbed to a far greater extent in a single solute system, but competition favored adsorption of DBS [Arbuckle and Romagnoli, 1980].

Arbuckle and Romagnoli (1980) used isotherm constants and chemical solubilities to predict the preferentially adsorbed compound in 22 bisolute

activated carbon systems. Solubility predicted the
preferred compound for 20 systems, while the
Freundlich K constant was correct for all systems.
The amount of nonpreferred material displaced was
in correlation with the absolute difference in K;
the greater the difference, the more material displaced. This information is useful in predicting
whether or not an organic will be concentrated in a
column´s effluent and to minimize carbon consumption by knowing the order the compounds leave the
column (important when removal of a specific component is desired).

Because many of the equilibrium theories and
theories for predicting competitive adsorption
currently in use are based on gas and vapor phase
adsorption, they have one major limitation: the
presence of solvent during solute adsorption is
ignored. Belfort and Altshuler have adapted the
solvophobic theory, a general thermodynamic treatment for describing the effect of various solvents
on reaction rate constants and equilibria, to adsorption [Miller, 1980b].

By using the same type of adsorbent and iden-
tical solution conditions, different workers have
attempted correlation of single-solute adsorptive
capacity with molecular weight, solubilty, functionality, and position of substitution. Although
several general trends are indicated, no one parameter has emerged in predicting the dominant effect
of adsorption. It is probably not reasonable to
expect a one-dimensional approach to provide a
consistent predictive correlation [Miller, 1980b].

Given single-solute or multisolute adsorption
equilibrium, a reasonably good description of multicomponent behavior may be obtained. There is a
need for an equilibrium adsorption theory which can
predict, without experimental observations, the
preferential adsorption of organic compounds onto
activated carbon from dilute aqueous solutions
[Miller, 1980a].

Recycle, Recovery, and Reuse Applications

Adsorption has been successfully applied for
recovery of numerous organics. Examples are listed
below:

Polysterene Adsorbent Resins
Polychlorinated Biphenyls
Ethylene Dichloride
Chloroform
Carbon Tetrachloride
Benzene, Toluene, Xylene
Phenols
Para-Nitrophenol
Phenoxy Acide Pesticides
Chlorinated Pesticides
Aliphatic Chlorinated
Hydrocarbons
Activated Carbon
Benzene

Toluene
Xylene
Ethylbenzene
Acrylic Monomers
Phenol
Acetic Acid
Aromatic Acids
Chlorinated Aromatics
Alcohols
Esters
Cresylics

Hydrocarbons
Tar Acids
Tar Bases

Industries in which recovery by adsorption has been applied include:

Pesticide Manufacturing
Organic Chemical Manufacturing
Petrochemical Industry
Paint Industry
Coke Plant

ADSORPTION OF GASES

In the recovery of organic gaseous air pollutants, activated carbon has been by far the most effective adsorbent used. However, the adsorptive capacity of any adsorbent is limited by:

- the surface area available for adsorption

- pore size and distribution

- temperature and pressure of operating conditions

- concentrations of influent organics

- desired recovery level (collection efficiency)

In the regeneration process, modification of systems and operating procedures used in solvent recovery have resulted in improved collection. However, this recovery is rarely optimized under given operating conditions. Vacuum stripping with steam has recently been shown to be a successful

method of recovering hydrocarbons [Kenson, 1979].
However, some problems arose due to corrosion effects on the adsorber vessel. Many of the materials recovered hydrolyze slightly to generate HCl
when exposed to chlorinated solvent. A better
understanding not only of physical properties but
also of chemical properties of both adsorbent and
adsorbate is necessary.

Theoretically, optimum adsorption is accomplished under the following conditions:

-        low base-line concentration of effluents

-        containment of breakthrough emissions

-        efficient recovery of desorbed organics

-        containment of organics exhausted during
         cooling and drying cycle

Since adsorption capacity is affected both by the
properties of the adsorbent and those of the adsorbate, and the conditions under which they are contacted, a better understanding of the properties of
both adsorbent and adsorbate would be beneficial.

Adsorbents and Their Properties

Each adsorbent has a preferential capacity for
each adsorbate based on adsorptive capacity and
affinity. The characteristics of selectivity is
also important for each adsorbent. Selectivity and
capacity can change the suitability of an adsorbent
for air pollution control usage, limiting some for
special conditions, while allowing broader application of others.

Chemical composition, ratio of surface area to
volume, pore size and distribution, and granule
(particle) size are among the most important characteristics of an adsorbent. The adsorption capacity is a function of these characteristics, and is
best defined with adsorption isotherms and/or adsorptive capacity data generated at operation conditions.

Depending on their chemical constitution and
pore size distribution, adsorbents can be classified into the four categoreis which are discussed
below.

1.   Chemically Reactive Adsorbents

These adsorbents can be impregnated with
chemically reactive compounds. The adsorption
operation by these adsorbents is highly
selective and tends to be irreversible. It is
an exothermic reaction which stops when all
active sites on the surface have reacted. The
surface is then covered with a unimolecular
layer of vapor.

Example: Soda lime (with or without
            activated carbon) is used in the
            chemisorption of ethanoic acid,
            acetonitrile, acrylonitrile, alkyl
            chloride, and vinyl propyl
            desulfide.

2.   Polar Adsorbents

The adsorption process is caused by van der
Waal's forces on a chemically non-reactive
surface, having polar properties. These
adsorbents are less selective and the process
is reversible. The heat released is about 10
kcal/g. mole. The dielectric constant and the
dipole moment of molecules characterize their
adsorption affinity. Affinity decreases with
decreasing dielectric constant and decreasing
dipole moment. However, it also decreases
with increasing van der Waal's force.

The adsorbing surface is saturated by several
molecule layers. At fixed vapor pressure, the
adsorption decreases with increasing
temperature. The application of these
adsorbents is mainly limited to the drying of
gases, due to their strong affinity for water.
Desorption generally occurs with regenerative
water steam followed by drying of adsorbent at
higher temperature.

Examples: Silica gel and activated alumina
             both have strong selectivity for
             polar compound (H_2O, olefins) and
             are generally used in separation and
             purification processes.

3.   Non-Polar Adsorbents

The most important adsorbing non-polar solid

is activated carbon, which is effective in
attracting non-polar molecules such as
hydrocarbons. Activated carbon is used to
remove hydrocarbons, odors and similar trace
contaminates from gas streams. Activated
carbon must be specially treated before use,
because adsorption of water vapor is very
sensitive to the presence of polar impurities.
Activated carbon usually contains S_iO_2, Al_2,
Fe_2O_3, NaOH, KOH and adsorbed oxygen.

4. Molecular Sieves

Molecular sieves having polar and non-polar
properties are available. They are effective
in adsorbing low molecular weight or
unsaturated hydrocarbons from dry air at low
concentrations. Adsorption of low molecular
weight saturated hydrocarbons (methane,
ethane) by activated carbon is more effective
than by molecular sieves, but too inefficient
in any case for effective pollution control.

Regeneration Operation

The adsorbent bed must be regenerated for
reuse after saturation has occurred. The recovery
of organic compounds generally occurs by stripping
them into easily condensable streams of gas
[Kensen, 1979]. The conventional methods of regeneration involve heated air, heated inert gas, or
steam, depending both on the adsorbent and the
adsorbate properties. In the case of use of noncondensable gas as a regenerative agent, the desorbed material can be disposed of in several ways
[US EPA, April 1973]. When using steam as a regenerating agent many pollutants cannot be economically recovered, due to their high steam to solvent
ratios. Adsorption/steam regeneration with solvent
recovery has been conventionally used in concentration above 500 mg/l. Using distillation on partially or totally soluble compounds is costly and,
moreover, disposal of this polluted steam condensate can constitute another form of environmental
pollution if discharged.

Some commercially available systems use hot
nitrogen or some other inert gases to desorb the
organics from the carbon. Vaporized organics condense to a liquid, and are reused in the process or
as a supplemental fuel. The inert gas is then

vented or reheated and recirculated to the
adsorber.

Regeneration can also be accomplished by
vacuum stripping [Kenson, 1979]. In a vacuum
system, pressure is reduced to a point below the
original partial pressure of adsorbed material.
Regeneration takes place rapidly at extremely high
concentrations. This method is used when there is
hydrolysis of the organics or formation of azeotrope with water, if the organics do not respond
sufficiently to low pressure steam regeneration, or
if they have concentration above combustion range.
Inert gas regeneration is also used under the above
conditons. A cooling and drying period is necessary before the bed can be reused for recovery.

Recent efforts of process engineers and equipment designers have been mainly oriented toward
improving the economics of solvent recovery operations. So far the principal areas of interest have
been [US EPA, April 1973]:

- Decrease in energy or fuel requirement
  for regeneration

- Reduction of pressure drop during
  adsorption phase

- Increase of adsorptive capacity by
  utilizing residual capacity of
  adsorption zone

- Improved methods for recovery of
  desorbed solvents.

The Adsorption Cycle

The adsorption cycle is a function of the
adsorbent life, the bed length, and the mass transfer zone. Vapor-phase activated carbon adsorption
has gained favor as a method of recovering valuable
solvents from industrial emission sources. The
adsorption is usually a batch operation with multiple beds. Upflow design is generally avoided because carbon particles can become entrained in the
exhaust at higher superficial velocities. The gas
stream is pretreated to remove solids (dust, lint),
liquids (droplets or aerosols), or vapor (high
inlet concentration or high boiling components),
since these can hamper performance. Pretreatment

with condensers is generally advantageous in reducing inlet concentrations. The following criteria are important for high removal efficiency
[Parmele, et al. 1979].

-    Low base-line effluent concentrations
     (usually less than 10 ppm).

-    Containment of breakthrough emissions.
     This is done by avoiding premature
     breakthrough or by sending material
     emitted during breakthrough to another
     vessel connected in series.

-    Efficient recovery of desorbed organics.
     This is usually achieved by condensing
     the vapors and recycling the
     noncondensables from the condenser back
     to the inlet of the online adsorber.

-    Containment of organics exhausted during
     the cooling and drying cycle. This
     cycle prepares the carbon bed for
     renewed service.

Application of Adsorption Operation in Industries:
Gases

     Adsorption is a particularly recommended and
useful technique when:

1.   The solvent (containment) has recovery
     value.

2.   The solvent is in extremely low
     concentrations.

3.   The solvent gas may be noncombustible or
     nonflammable (below 25 percent of its
     lower explosive limit).

Some applications of gas adsorption are:

-    surface coating (fabric or filter)
-    film casting
-    metal degreasing
-    dry cleaning
-    printing
-    rendering
-    food processing
-    chemical processing

- paint spray booth
- bake ovens

Some of the principal adsorbents used are:

- activated carbon: hydrocarbons, odors
- silica gel: dehydration of gases
- activated alumina (aluminum oxides): dehydration
- molecular sieve (synthetic zeolites): SO_2 , NO_x , Hg

REFERENCES

Arbuckle, W.B. and Romagnoli, R.J.   "Prediction of the Preferentially Adsorbed Compound in Bisolute Column Studies." AIChE Symposium Series, No. 197 Vol. 76:77, 1980.

Baker, C.D., Clark, E.W., Jessering, W.V. and Huether, C.H. "Recovering Para-Cresol from Process Effluent." Chemical Engineering Progress, 69(8):77, 1973.

Baker, E.B., Anti-Pollution By-Product from Phenolic Contaminants, U.S. Pat. 3, 907, 524, September 23, 1974.

Cheremisinoff, P., and Habib, Y.H. "Cadmium, Chromium, Lead Mercury. A Primary Account for Water Pollution. Part 2. Removal Techniques." Water Sew. Works. 46-51, 1972.

Cheremisinoff, P. and Morressi, A.C. "Carbon Absorption." Poll. Eng. 6(8): 66-68, 1974.

Cheremisinoff, P. and Morresi, A.C., "Carbon Adsorption Applications." Carbon Adsorption Handbook. (Eds.) Cheremisnoff, P.N. and Ellenbusch, F., Ann Arbor Science Publishers, Inc., Ann Arbor, Michigan, 1978.

E. Crook, R.P. McDonnell, J.T. McNulty, "Removal and Recovery of Phenols from Industrial Waste Effluents with Amberlite XAD Polymeric Adsorbents," Industrial and Engineering Chemical Products Research and Development, 14(2):113, June, 1975.

F.A. DiGiano, "Toward a Better Understanding of the Practice of Adsorption." AIChE Symposium Series, No. 197, Vol. 76:86, 1980.

M.A. El-Dib, and Badawy, M. "Adsorption of Soluble Aromatic Hydrocarbons on Granular Activated Carbon." Water Research, (GB) 13:225, 1979.

Fox, C.R. "Plant Uses Prove Phenol Recovery with Resins." Hydrocarbon Processing, 57(11):269, 1978.

Fox, C.R. "Removing Toxic Organics from Wastewater." Chemical Engineering Progress, 75(8):70, 1979.

Fox, R.D. and Himmelstein, K.J. "Recovery of Destruction-New Developments for Industrial Wastes." p. 445 Proc. of the 29th Industrial Waste Conference, May, 1974.

Geankoplis, C.J. Transport Processes and Unit Operations. Allen and Bacon, Inc., Boston, 1978.

Hager, D.G. "Industrial Wastewater Treatment by Granular Activated Carbon." Ind. Waste Eng. 14-28, 1974.

Haines, H.W. Jr., Vapor Recovery Method for Controlling Air Pollution, U.S. Pat. 3, 907, 524, September 23, 1974.

Hasler, J.W. Purification with Activated Carbon. Chemical Publishing Company, Inc., New York, 1974.

Himmelstein, K.J. Removal of Acetic Acid from Wastewaters." p. 677 Proc. of the 29th Industrial Waste Conference, May, 1974.

Himmelstein, K.J., Fox, R.D., and Winter, T.H. "In-Place Regeneration of Activated Carbon." Chemical Engineering Progress, 69 (11):65, November 1973.

Hutchins, R.A. "Economic Factors in Granular Carbon Thermal Regeneration." Chemical Engineering Progress, 69(11):48, November 1973.

Jensin, R.A.   "Semi-Continuous Activated Carbon
        Systems for Wastewater Treatment." AIChE
        Symposium Series, No. 197, Vol 76:77, 1980.

Kenson, R.E.   "Carbon Adsorption of Hydrocarbon
        Emissions Using Vacuum Stripping." Pollut.
        Eng. V11 N7 Jul. 1979 (p. 38-40).

Kim, B.R., Snoeyink, V.L. and Saunders, F.M.
        "Adsorption of Organic Compounds by Synthetic
        Resins." Journal of Water Pollution Control
        Federation, 48(1):120, 1976.

Loop, G.C.   "Refinery Effluent Water Treatment
        Plant Using Activated Carbon. U.S. EPA, EPA
        600/2-75-020, 1975.

Matsumato, K., S. Kurisi, T. Oyamoto, Development
        of Process of Full Recovery by Thermal
        Decomposition of Waste Plastics, Mitsubishi
        Heavy Ind., Ltd., 1st Int. Conf., Conversion
        of Refuse to Energy, Montisux, Switzerland,
        1975.

Mattson, J.S. and H.B. Mank, Jr. Activated Carbon.
        Marl Dekker, Inc., New York, 1971.

Miller, S., "Adsorption on Carbon: Theoretical
        Considerations." Environmental Science and
        Technology, 14(8):910, August, 1980a.

Miller, S. "Adsorption on Carbon: Solvent Effects
        on Adsorption." Environmental Science and
        Technology, 14(9):1037, September 1980b.

Netzer, A., Wilkinson, P. and Beszedits, S.
        "Removal of Tracemetals from Wastewater by
        Treatment with Lime and Discarded Automotive
        Tires." Water Res. 8():813-817, 1974.

Nichols, R.A., Hydrocarbon-Vapor Recovery, Chem.
        Eng. 80(6):8592, March 5, 1973.

Oda, T., K. Chibo and H. Hosakawa, Thermal
        Composition of Plastic Waste Containing Poly-
        Vinyl Chloride, 1st Int. Conf. Conversion of
        Refuse to Energy, 1975, Montieus, Switzerland,
        1975.

Parmele, C.D., O'Connel, W.L. and Basdekis, H.S.
    "Vapor-Phase Adsorption Cuts Pollution,
    Recovers Solvent." Chem. Eng. Series: B6
    Issue: 2B Dec. 1979 (p. 58-70).

Ross, R.D., Selection of Equipment for Gaseous
    Waste Disposal in Air Pollution and Industry,
    Ch. 9 p. 422, 478. N.Y., Van Nostrand
    Reinhold, 1972.

Smithson, G.R., Jr. "An Investigation of
    Techniques for the Removal of Chromium from
    Electroplating Wastes." Water Poll. Control
    Res. Ser. No. #12010 EIE 03/71, 1971.

Snoeyink, V.L., Weber, W.J., and Mark, Jr., H.B.
    "Sorption of Phenol and Nitrophenol by Active
    Carbon." Environmental Science and
    Technology, 3(10):918, October, 1969.

Stevens, B.W. and Kerner, J.W. "Recovering Organic
    Materials from Wastewater." Chemical
    Engineering, p. 84, February 3, 1975.

U.S. EPA. "Process Design Manual for Carbon
    Adsorption." Tech. Transfer Series, 1971a.

U.S. EPA. "An Investigation of Techniques for
    Removing Cyanide from Electroplating Wastes."
    Water Poll. Cont. Re. Ser. No. #12010 EIE.
    11/71, 1971b.

U.S. EPA Package Sorption Device System Study,
    Document No. PB-22-138, April 1973.

Weber, W.J., Jr. Physiochemical Processes for Water
    Quality Control. Wiley-Interscience, New
    York, 1972.

SUPPLEMENTAL REFERENCES

Abel, W.T., Shultz, F.G. and Langdon, P.F.
    "Removal of Hydrogen Sulfide from Hot Producer
    Gas by Solid Absorbents." Bureau of Mines,
    Morgantown, W. Va., Morgantown Energy Research
    Center, RI 7949, 28 p., 1974.

Abrams, I.M. "Removal of Organics from Water by
    Synthetic Resinous Adsorbents." Chemical
    Engineering Progress, 65(97):106, 1969.

Abrams, I.M. "Macroporous Condensate Resins as Adsorbents." Industrial and Engineering Chemical Products and Research Development, 14(2):108, June, 1975.

Ammons, R.D., Dougharty, N.A. and Smith, J.M. "Adsorption of Methyl-Mercurik Chloride on Activated Carbon Rate and Equilibrium Data." Ind. Eng. Chem. Fund. 16(2):263-269, 1977.

APCA Proc. on State of the Art of Odor Control Technology II Specialty Conference, March 1977, Odor Control by Adsorption by E.D. Ermence.

Argaman, Y., and Weddle, C.L. "Fate of Heavy Metals in Physical-Chemical Treatment Processes." AIChE Symp. Ser. 70(136):400-414, 1973.

Badhwar, K. "Chlorine Recovery with Aqueous Hydrochloric Acid." U.S. Pat. 3, 881, 893, 5 p., May 6, 1975.

Balakrishnam, S. and Rickles, R.N. "By-Product Recovery and Air Pollution Control." Reprint, APCA., Pittsburgh, Pa., 191, 1972.

Barneby, H.L., "Activated Charcoal in the Petrochemical Industry." Chemical Engineering Progress, 67(11):49, November, 1971.

Berkowitz, J.B., et al. "Physical, Chemical, and Biologic Treatment Techniques for Industrial Wastes." NTIS #PB275054, November, 1976.

Bernardin, F.E. "Detoxification of Cyanide by Adsorption and Catalytic Oxidation on Granular Activated Carbon." Proc. 4th Mid-Atl. Ind. Waste Conf. pp. 203-228, 1971.

Bottoms, R.R. "Process for Separating Acidic Gases." U.S. Patent, 1, 834, 016, 5 p., December 1, 1931.

Bodoni, D. "Process and Equipment for the Recovery of Usable Components from Flue Gases." Text in German, Austrian Pat. 174, 896, 4 p., May 11, 1953.

"Can H S Stack be Easily Recovered." Can. Chem. Process, 53(8):56-57, August, 1969.

Cheremisinoff, P. Control of Gaseous Air Pollutants, Pollut. Eng. VB N5 May 1976 (p. 30-36).

Chien, H., "Sulfuric Acid Plant Tail Gas Absorption Experiment." Text in Japanese, Kung Chen (Eng. J.), 45(3):924, 1972.

Davies, R.A., Kaempf, H.J., and Clemens, M.M. Removal of Organic Material by Adsorption on Activated Carbon." Chemistry and Industry, p. 287, September 1, 1973.

Downer, W. "Ammonia Absorption: Refrigeration Selected for Gasoline Plant." Refining Eng., Vol. 29: C25 to C30, July, 1957.

Drechsel, H., et al., "Production of Sulfur Trioxide and Sulfuric Acid." U.S. Pat. 3, 525, 586, 7 p., August 25, 1970.

Erskine, D.B., and Schuliger, W.G. "Graphical Method to Determine the Performance of Activated Carbon Processes for Liquids." AIChE Symp. Ser. 68(124):185-190, 1971.

Ford, C.T., and Boyer, Jr., J.F. "Treatment of Ferrous Acid Mine Drainage with Activated Carbon." EPA Report. EPA-R2-73-150, 1973.

Fox, R.D., Keller, R.T. and Pinamont, C.J. "Recondition and Reuse or Organically Contaminated Waste Sodium Chloride Brises." U.S. EPA Report No. EPA-R2-73-200, 1973.

Fukui, S., et al. "Method of Eliminating Nitrogen Oxides from Exhaust Gas." Text in Japanese. Japan Pat. 47, 10843, 5 p., March 31, 1972.

Ganz, S.N. et al. "Removal of Nitrogen Oxides, Sulfur Dioxide, Mist and Sulfuric Acid Spray from Exhaust Gas by Peatalkali Sorbent Under Production Conditions" J. Appl. Chem. USSR (English Translation from Russian of: Zh. Prikl. Khim). 41(4):700-704, April 1968.

Garten, V.A. and D.E. Weiss. "Ion and Electron
    Exchange properties of Activated Carbon in
    Relation to its Behavior as a Catalyst and
    Absorbent." Rev. Pure Appl. Chem. 7:69, 1957.

George, A.D. and Chaudhuri, M. "Removal of Iron
    from Ground Water by Filtration Through Coal."
    J. Am. Water Works Assoc. 69:305, 1977.

Gradjacques, B. Carbon Adsorption Can Provide Air
    Pollution Control with Savings, Pollut. Eng.
    V9 N8 Aug. 1977 (p. 28-31).

Hager, D.G., "Industrial Wastewater Treatment by
    Granular Activated Carbon," Industrial
    Engineering, February, 1974.

Harvin, R.L. "Modern Design Solvent Recovery
    Plant." Presented at AIChE Symp. Series 126
    V68, 1972, p. 302(5).

Helsel, R.W. "A New Process for Recovering Acetic
    Acid from Dilute Aqueous Waste Streams."
    p. 1059, Proc. of the 31st Industrial Waste
    Conference, May, 1976.

Hinricks, R.L., Snoeyink, V.L. "Sorption of
    Benzenesulfonates by Weak Base Anion Exchange
    Resins." Water Research, 10:79, 1976.

Hiroshi, S. and Nakamoto, Y. "Absorption of SO
    Gas." Translated from Japanese. Franklin
    Inst. Research Labs, Philadelphia, PA.,
    Science Info. Services, 8 p., October 1969.

Hixson, A.W. and Miller, R. "Recovery of Acidic
    Gases." U.S. Patent 2, 449, 537, 4 p.,
    September 21, 1948.

Hoar, F.J. "System for Recovering Relief Gases
    from a Sulphite Pulp Digester." U.S. Pat. 3,
    313, 680, 5 p., April 11, 1967.

Hopwood, A.P. "Protein Recovery." Effluent and
    Water Treatment Journal (GB) 8:333, 1978.

Huang, J.C. and Steffens, C.T. "Competitive
    Adsorption of Organic Materials by Activated
    Carbon," p. 107, Proc. of the 31st Industrial
    Waste Conference, 1976.

Humenick, M.J., Jr. and Schnoor, J.L. "Improving Mercury (II) Removal by Activated Carbon." J. Am. Soc. Civil Environ. Eng. Div. 100(6):1249-1262, 1974.

Huntington, R.L. "Flue Gas Recovery Method and Apparatus." U.S. Pat. 3, 733, 777, 9 p., May 22, 1973.

Huntington, R. "Multiple Compartment Packed Bed Absorber-Desorber Heat Exchanger and Method." U.S. Pat. 3, 791, 102, 9 p., February 12, 1974.

Illinicz, J. and L. Porowski, "Utilization of Steel Furnace Dusts by the Method of Chemisorption." Air Conserv. 3(5):13-17, 1969.

Jain, J.S. and Snoeyink, V.L. "Competitive Adsorption from Bisolute Systems on Active Carbon." Journal of Water Pollution Control Federation, 45:2463, 1973.

Jones, H.R. "Removal of Mercury From Gases." In: Mercury Pollution Control, Noyes Data Corp., Park Ridge, N.J., 15 p., 1971.

Jones, W.J. and Ross, R.A. "The Sorption of Sulfur Dioxide on Silica Gel." J. Chem Soc. A. 1967: 1021-1026, 1967.

Kafarov, V.V., et al. "Study of the Dynamism of the Non-Isothermal Absorption Process in Industrial Packed Absorber with Recycling." Text in Russian. Khim. Pro., (Moscow), No. 2: 5759, 1972.

Kakabadze, V.M. and Kakabadze, I.L. "Absorption of Nitrogen(II) Oxide Gas by the Dry Method with Simultaneous Production of Fertilizer." Text in Russian, Soobsheh, Adad, Hank Gruz, SSR, 18(5): 549-556, 1957.

Kasal, T., "Konox Process Removes H_2S." J. Hydrocarbon Processing, February 1975, V54, N2, p. 93(3).

Kato, S. "Deodorizing Process by Chemical Absorption, and its Problems." Text in Japanese. Akusu no Keukyo (Odor Research J. Japan), 3(11):25-40, May, 1973.

Kattan, A. and Gwyn, J.E. "Vapor Recovery and Disposal System." U.S. Pat. 3, 097, 193, 6 p., July, 1975.

Kauase, B., Jogima, T. and Otani, K. "Method for Removing Mercury Vapors Contained in Gas." Text in Japanese, Japan Pat. Sho 4843257, 2 p., December 18, 1973.

Kawabata, N. and Ohira, K. "Removal and Recovery of Organic Pollutants from Aquatic Environment-I. Vinylpyridine-Divenylbenzene Copolymer as a Polymeric Adsorbent for Removal and Recovery of Phenol from Aqueous Solution" Environmental Science and Technology, 13(11):1396, November, 1979.

Kennaway, T., Wood, C.W., and Box, P.L. "A New Development in the Production of By-Product Ammonium Sulphate." Gas World, Vol. 143:49-58, March 3, 1958.

Kennedy, D.C. "Treatment of Effluent from Manufacture of Chlorinated Pesticides with a Synthetic Polymeric Adsorbent." Environmental Science and Technology, 2(1) 154, 1968.

Kostyuchenko, P.I., Tarkovskaya, F.A., Kononchuk, T.I., Kovalenko, T.G., and Glushankova, Z.L. "The Mechanism of Selective Sorption of Traces of Vanadium Ion by Activated Charcoal." In: Adsorption and Adsorbent. John Wiley & Sons, Inc. p. 37, 1973.

Kunz, R.G., Giannelli, J.F., and Stensel, H.D. "Vanadium Removal from Industrial Wastewater." J. Water Poll. Control Fed., 48(4):762, 1976.

Lefrancois, P.A. and Barclay, K.M. "Purification of Waste Gases." U.S. Pat. 3, 671, 185, 6 p., June 20, 1972

Leiderback, T.A. "Reducing Chlorine Loss in Electrolysis Plant." J. Chem. Engr. Progress, March, 1974, V70 N3, pg. 4(s).

Lin, Y.H. and Lawson, J.R. "Treatment of Oily and Metal-Containing Wastewater." Pollut. Eng. 5(11):45-48, 1973.

Loven, A.W. Perspectives on Carbon Regeneration.
    Chemical Engineering Progress, 69(11):56,
    November, 1973.

Loven, A.W. "Activated Carbon Regeneration
    Perspectives." AIChE Symp. Ser. 70(144):
    285-295, 1974.

Lowell, P.S. and Parson, T.B. "A Theoretical Study
    of NO_x Absorption Using Alkaline and Dry
    Sorbents. Vol. I." (Final Report). Radian
    Corp., Austin, Tex., Office of Air Programs
    Contract EHSD 715, APTD1162, 100 p., December
    31, 1971.

Malhur, S.B., et al., "Recovery of Mercury From
    Effluents." Chem. Age. India, 23(4):284-290,
    April 1972.

Marchenko, Yu G. and Nouikov, V.E. "Determining
    the Minimum Losses of Benzol Hydrocarbons in
    the Final Coke-Oven." Gas. Coke Chem. (USSR)
    (Translated from Russian), No. 11:30-32, 1970.

Martinola, F. and Richter, A. "Macropourous Resins
    as Organic Scavengers," Industrial Water
    Engineering, 8:22, 1971.

Maryland, B.J. and Heinz, R.C. "Continuous
    Catalytic Absorption for Nitrogen Oxides
    Emission Control." AIChE Symp. Ser. V70 N137,
    (p. 2387), 1974.

Molvar, A.E., Rodman, C.A., and Shunney, E.L.
    "Treating Textile Wastes with Activated
    Carbon." Textile Che. Colorist. 2(16):286-290,
    1970.

Moore, R.H. "Investigation of a Process for
    Removal of Copper From Seawater Desalination
    Plant Effluent Using Carbon Sorbates." U.S.
    DOI Report #651, 1971.

Musty, P.R. and Nickless, G. Use of Amberlite
    XAD-4 for Extraction and Recovery of Chlori-
    nated Insecticides and PCB's from Water.
    Journal of Chromatography, 89:185, 1974.

Nelson, F., Phillips, H.O., and Kraus, K.A.
"Adsorption of Inorganic Materials on
Activated Carbon." <u>Proc. 29th Ind. Waste
Conf</u>., PurdueUniversity, pp. 1076-1090,
1974.

Oehme, C. and Martinola, F. <u>"Removal of Organic
Matter from Water by Resinous Adsorbents</u>,"
Chemistry and Industry, p. 823, September 1,
1973.

Oloman, C., Nurray, F.E. and Risk, J.B. "The
Selective Absorption of Hydrogen Sulfide From
Stack Gas." Palp Paper Mag. Con. (Quebec),
1969: 69-74, December 5, 1969.

Ozerskii, Yu G., <u>et al</u>. "Recovery of Phenols and
Hydrogen Sulphide From Wastes Discharged to
Atmosphere." <u>Coke</u> <u>Chem</u>. (USSR), (Translated
from Russian), Vol. 6:41-44, 1969.

Parkes, D.W. and Evans, R.B. "Absorption of Acidic
Gases." U.S. Patent 2, 106, 435, 3 p.,
January 25, 1938.

Parmelle, C.S. and Fox, R.D. "Reuse Comes Out
Ahead." <u>Water</u> <u>and</u> <u>Wastes</u> <u>Engineering</u>,
9(11):10, 1972.

Perrotte, D.E., and Rodman, C.A. "Factors Involved
with Biological Regeneration of Activated
Carbon." <u>AIChE Symp. Ser</u>. 70(144):316-325,
1974.

Ranke, Gerhard, <u>et al</u>. "Procedure and Equipment
for the Production of Hydrogen and Carbon
Monoxide." Text in German, W. Ger. Pat.
Appl., 2, 025, 763, 18 p., May 26, 1970.

Rey, G., Dick, M., and DesRosiers, P. <u>EPA´s R&D
Program for Activated Carbon</u>," Chemical
Engineering Progress, 69(11):45, November,
1973.

Roots, D.C. "The Elimination of Atmospheric
Pollution From Stoving Ovens." <u>Ind. Finish.
Surf. Coatings</u>, 26(312):45, June, 1974.

R.S. Kerr Environmental Research Laboratory,
Activated Carbon Treatment of Industrial
Wastewaters: Selected Technical Papers.
EPA/600/2-79/177, August, 1979.

Reimers, R.S. and Englande, A.J. "A Quick Method
for Evaluating the Suitability of Activated
Carbon Adsorption for Wastewater," Proc. of
the 31st Industrial Waste Conference, 1976.

Saburo, Fukui, et al. "Chemical Recovery Process
of High Sulfidity Spent Cooking Liquids."
Text in Japanese. Kamipa G. Kyoshi, 24(10):
515-522, October, 1970.

Schmitt, Karl, et al. "Nitrogen Oxide Conversion."
U.S. Pat. 3, 453, 071, 7 p., July 1, 1969.

Schwanecke, R. "Waste Gas Cleaning Through
Combustion of Nitrogen Oxides." Text in
German. Zentr. Arbeiksmed. Arbeitsschutz,
19(9): 262-264, 1969.

Shanmugasundaram, S., et al. "Pollution From
Fertilizer Plants: Treatment of Waste Waters
Removal of H_2S and CO_2." Chem. Age India, V26
N4 (p. 279-283), April, 1975.

Skovronek, H.S., Dick, M., and DesRosiers, P.S.
"Selected Uses of Activated Carbon for
Industrial Wastewater Pollution Control,"
Industrial Water Engineering, p. 6 May/June,
1977.

Steineke, F. "Method of Recovering Flourine From
Waste Gases." U.S. Pat. 3, 812, 852, 3 p.,
June 4, 1974.

Sultzman, R.S. and Hunt, Jr., E.B. "A Photometric
Analyzer System for Monitoring and Control of
the H_2S/SO_2 Ration in Sulfur Recovery Plants."
ISA (Quatr. Sol. Am) Truss., 12(2):103-107,
1973.

Sunney, E.L., Perbotti, A.E., and Rodman, C.A.
"Decolorization of Carpet Yarn Dye Wastewater."
American Dyestuff Reporter, 1971.

Synder, A.J., and Alsbaugh, T.A. "Catalyzed Bio-
Oxidation and Tertiary Treatment of Integrated
Textile Wastewaters." US EPA, 1974.

Takahaski, T. "NO Removal Techniques for Fixed
    Combustion System Flue Gases." Text in
    Japanese, Maruzen Sekiju Giho, No. 18:19,
    1973.

Thiem, L., Badorek, D., and O'Connor, J.T.
    "Removal of Mercury from Drinking Water Using
    Activated Carbon." J. Am. Water Works Assoc.
    68(8):447-451, 1976.

Timson, G.F. and Helein, J. "How Amoco Controls
    H_2S Clean-up." Hydrocarbon Process, 53(1):
    115-116, January, 1974.

Trebor Busby, H.G. and Darby, K. "Efficiency of
    Electrostatic Precipitators as Affected by the
    Properties and Combustion of Coal." J. Inst.
    Fuel (London), Vol. 36:184-197, May 1963.

U.S. EPA Process Design Manual for Carbon
    Adsorption, EPA Technology Transfer Manual,
    1973.

Valtsw, V.N., et al. Recovery of Hydrogen Iodide
    from the Gas Stream in the Venturi Scrubber in
    the Oxidative Dehydrogenation of Hydrocarbons.
    Text in Russian. Khim, Prow (Moscow), No.
    11:26-28, 1969.

Vanstone, G.R. and Gilmore, D.G. "Application of
    Granular Activated Carbon to Industrial Waste-
    water Treatment," Conference on Complete
    Water Reuse, 1973.

Weber, W.J., Jr. "Sorption from Solutions by Porous
    Carbon." In: Principles and Applications of
    Water Chemistry. S.D. Faust, and J.V. Huntor
    (eds.). John Wiley & Sons, Inc. 1967.

Welib, R.G. "Isolating Organic Water Pollutants:
    XAD Resins, Urethane, Foams, Solvent Extrac-
    tion," EPA/660/4-75-003, June, 1975.

Zaytseb, V.A., et al. "Waste Products From the
    Production of Phosphate Fertilizers as
    Possible Raw Material for Fluorine
    Production." Soviet Chem., Ind., No. 547-550,
    August, 1971.

CHAPTER 6

MOLECULAR SEPARATION

INTRODUCTION

This chapter examines recycle, recovery, and reuse applications of several molecular separations processes. The processes included in this chapter are:

1. Reverse Osmosis
2. Ion Exchange
3. Ultrafiltration

Ultrafiltration and reverse osmosis are commonly referred to as membrane processes. A membrane is defined as a phase which acts as a barrier to flow of molecular or ionic species between other phases that it separates. On the other hand, ion exchange can be considered a chemi-sorption process [Sundstrom and Klei, 1979]. In ion exchange, there is a reversible interchange of ions between a liquid and solid (the transfer of ions between phases occurs at the solids surface), where there are no permanent changes in the structure of the solid.

This chapter is divided into two major sections. The first section of the chapter examines recycle, recovery, and reuse applications for organic compounds as related to the two membrane processes listed above. The second section of the chapter is concerned with recycle, recovery, and reuse applications for inorganic compounds as related to the ion exchange and reverse osmosis processes.

ORGANIC COMPOUNDS

Membrane Processes

Ultrafiltration and reverse osmosis are hydraulic pressure activated processes capable of

separating solution components largely on the basis
of molecular size and shape, and involve neither a
phase change nor interphase mass transfer. By
proper membrane selection, it is possible to concentrate, purify, and fractionate many components
of a solution. These processes are particularly
attractive due to the fact that the sole energy
requirement is the compression energy of the feed
liquid.

Reverse osmosis generally applies to the separation of low-molecular weight solutes such as
salts, sugars, and simple acids from their solvent.
The driving pressure for efficient separation must
exceed the osmotic pressure of the solute in solution. This may require pressures of 500 to 2,000
psi. Ultrafiltration is the term used for separation of higher molecular weight solutes such as
proteins, starch, natural gums, and other complex
organic compounds from their solvents, as well as
colloidally dispersed substances such as clays,
pigments, minerals, latex particles, and microorganisms. Osmotic pressure of the solute in these
systems is usually negligible, and typical operating pressures range between 5 and 100 psi. Although the processes are related, ultrafiltration
does not require the high operating pressures which
are needed to overcome the high osmotic pressure
differential across a reverse osmosis membrane
[Porter, 1972]. Both systems, however, are worth
consideration in treating waste streams to produce
a concentrate suitable for reuse.

Membrane Characteristics

The membranes for both ultrafiltration (U.F.)
and reverse osmosis (R.O.) can be made from various
synthetic or natural polymeric materials, ranging
from hydrophilic polymers such as cellulose, to
hydrophobic materials such as fluorinated polymers.
Polyarylsulfonates and inorganic materials have
also been developed, to deal with high temperatures
and pH needs [Berkonitz, et al., 1976].

The use of cellulose acetate membranes or
their derivatives imposes some limitations. Operating temperatures must be restricted to less than
110°F to avoid hydrolysis of the cellulose acetate.
Cellulose acetate membranes normally are limited to
operating within the pH range of 3 to 8. Strongly
acid or alkaline solutions will cause rapid

membrane degradation [Ruver, 1973]. Certain non-cellulosic membranes have temperature capability of 230°F and pH capability between 1 and 12 [Porter, 1972].

Membranes having an anisotropic structure, with pores roughly conical in shape, are not subject to internal fouling problems. The smallest pore diameter remains at the membrane surface. A solute which finds its way into the membrane is likely to pass through to the other side of the membrane, since pore diameter increases from entrance to exit [Porter, 1973].

Ohya, et al. (1979) studied the effect of evaporation period at casting stage on the water flux and membrane structure of a cellulose acetate butyrate membrane. With short evaporation time, membranes with large fingerlike cavities were formed. No holes were formed on the upper surface. Water flux properties were low. Upon long evaporation periods, a solid structure was formed. These membranes tended to have high flux properties. No explanation was offered for the relationship between membrane structure and its characteristics.

Membrane Rejection and Flux

According to Spatz (1973), the rejection of organics is based on a sieve mechanism related to the size and shape of the organic molecule. In the case of organics which act like salts, a combination of rejection mechanisms may occur, since salts are rejected because of physio-chemical reaction with the membrane surface. Sourirajan and his coworkers have studied organic separation by cellulose acetate membranes, and found solute separation to be dependent on its ability to form hydrogen bonding with the membrane materials [Chian and Fang, 1977].

Chian and Fang (1977) in attempting to establish generalized criteria conducted with five membrane types and dozens of organic compounds, and found that the physio-chemical criteria and pressure effects on solute separation established with the cellulose acetate membrane are generally valid with other commercially feasible membranes. For instance, for membranes having appropriate surface structure, pore size, etc., the more nonpolar the membrane, the better the solute separation will be,

particularly for low molecular weight polar organic
solutes. The choice of nonpolar membrane materials, however, may produce a decrease in water
flux.

     Once a membrane is selected which provides the
desired removal of solute, the remaining important
operating parameter is membrane permeability or
flux. Initially, membrane resistance controls
flux. However, when the solute begins to build up
along the membrane surface, a polarized layer or
gel forms. When the gel concentration exceeds a
critical level, gel resistance controls mass transfer [Nelson, 1973].

     In the presence of large surfactant concentra-
tions, almost instantaneous fouling can take place,
a condition to which only low surface energy polymers are immune. Fouling can be controlled to a
degree by high recirculation velocities, proper
membrane configuration, and control of process
fluid composition [Mir, et al. 1977].

     Fenton-May and his coworkers [Fenton-May, et
al. 1972] found that at low flow rates, the primary
resistance to membrane transport is offered by a
hydrodynamic boundary layer produced by polarization, and that mass transfer is controlled by membrane resistance and gel precipitate at higher flow
rates. For both reverse osmosis and ultrafiltration, Fenton-May and his coworkers found an increase in permeate flux of 40 percent for every $20\,^{\circ}F$
increase in feed temperature.

     In studying pressure effects, they found a
completely linear relationship between applied
hydrostatic pressure and membrane pure water flux,
up to a pressure of about 450 psi, indicating that,
within this range, resistance of the membrane itself to mass transport is pressure independent.
When a whey or skim milk feed was used, however, an
increase in operating pressure increased solute
rejection, a result of an increase in resistance to
solute transport of the protein gel layer.

     Since membrane flux rates are, for the most
part, limited by concentration polarization or gel
resistance, various techniques have been applied to
reduce this problem, generally by cross-flow fluid
management techniques. It has been reported that
by operating above a critical cross-flow velocity,

gel resistance is minimized and flux increased with increasing pressure. An economic optimum must be reached between power costs necessary to maintain the desired cross-flow velocity and shear effects.

Prediction of Binary-Solute Behavior

Fels (1972) studied the permeation and separation of binary organic mixtures using polyethylene membranes, and found that ideal behavior was not exhibited. The deviation from ideal behavior increased as the difference in interaction behavior between the liquids and the polymer (as measured by a solubility parameter difference) increased.

Statement of Limiting Technology

1. Membrane Characteristics

    •   The relationship between membrane structures and their rejection and flux properties are not fully understood.

    •   A need exists for better membranes which will have longer life under the varying operating conditions of temperature, pressure, pH, and flow rate, which are less prone to fouling and chemical degradation, have higher water flux rates, and good solute rejection characteristics.

2. Membrane Rejection and Flux

    •   The membrane properties of flux and rejection rarely control process performance. Performance is usually limited by concentration polarization or gel formation at the membrane surface, and is less efficient than membrane controlled performance. The answer to the problem may be found in the study of one of three areas: improved hydrodynamics resulting in higher shear forces; the interaction of gel layers, or dynamic membranes, with the operating membrane, or the interaction between the membrane polymer and feed.

- Little is known about the possibility of reducing the influence of undesirable dynamic membranes formed from feed constituents. It may be possible to purposely add species capable of forming gel layers, or secondary membranes, which may enhance membrane performance.

Resource, Recovery, and Reuse Applications

- Production of protein concentrates and isolates from soy whey.

- Concentration and fractionation of cheese whey and skim milk.

- Oil-emulsion concentration.

- Polyvinyl alcohol recovery and concentration.

- Recovery of protein and pectin from sugar beet wastes.

- Recovery of lignin and ligno-sulfonates from pulp and paper extraction effluent.

- Recovery of polyglycols from polyurethane production processes.

- Recovery of chocolate wastes.

INORGANIC COMPOUNDS

This section examines recycle, recovery, and reuse applications for inorganic compounds as related to the ion exchange and reverse osmosis processes.

Ion Exchange

Ion exchange is a process in which ions held by electrostatic forces to charged functional groups on the surface of a solid, are exchanged for ions of similar charge in a solution in which the solid is immersed [Weber, 1972]. Basically, in this process, the ion exchanger is contacted with the solution containing the ion to be removed until the active sites on the exchanger are partially or completely "exhausted" by that ion. The exchanger may then be contacted with a sufficiently concentrated solution of the ion originally associated with it to "regenerate" it by thermodynamic forces back to its original form.

The exchange reactions during "exhaustion" and "regeneration" can be represented by the following two reactions:

Exhaustion: $M^+X^- + R^-X^+ \rightleftarrows R^-M^+ + N^+ + X^-$ (6.1)

Regeneration: $R^-M^+ + N^+$ (high concentration)
$$\rightarrow R^-N^+ + M^+$$ (6.2)

In the first reaction, a cation exchange material designated R^-, having a cation N^+ associated with it reacts with a solution of an electrolyte (M^+X^-). In the regeneration reaction, concentration of N^+ is higher than in the exhaustion reaction and the solution volume for regeneration is smaller, thereby making concentration of M^+ higher in the regeneration solution.

Regeneration of ion exchanger may also involve a third ion (for example a hydrogen ion in cation exchange) to give the corresponding form of the exchanger which can then be converted to the desired salt as shown below:

$$R^-M^+ + H^+ \rightarrow R^-H^+ + M^+$$ (6.3)

$$R^-H^+ + N^+OH^- \rightarrow R^-N^+ + H_2O$$ (6.4)

Such a sequence might be desirable in a case where H^+ was a much more effective regenerant than N^+ but where the release of H^+ into the solution being purified will be undesirable [Arthur D. Little, Inc., 1976].

Thus, during the process of ion exchange, undesirable ions from a stream are removed and transferred at a higher concentration to another aqueous stream. The phenomenon of ion-exchange is known to occur with a number of natural solids (for example, soil, humus, metallic oxides, etc.) as well as synthetic resins, which are presently used for most ion exchange applications.

Types of Ion Exchange Operations

Ion exchange operations are basically batch type but may be used on a semi-continuous basis. There are four operations carried out in a complete cycle, namely, service (to exhaustion), backwash, regeneration, and rinse. There are three principal operating models in use today: concurrent fixed-bed, counter-current fixed bed, and continuous counter-current. Most ion exchange installations in use today are of the fixed bed type with counter-current operation becoming more popular, especially for removal of traces of hazardous species from the waste stream prior to reuse or discharge [Arthur D. Little, Inc., 1976].

Ion Exchange: State-of-the-Art

One major application of the ion exchange process is in the metal-finishing industry for the recovery of chromic acid from rinsewaters. The acidity of a chrome bath of average use is so high (pH 0.5 or less) that the use of any presently known cation exchange resin for direct removal of cations from the bath is impossible. Also, the high oxidative potential of the bath results in rapid resin degradation of almost all available cation exchange resins [Raman and Carlson, 1977].

In a typical chromic acid recovery system, the acid rinse stream is first neutralized to a pH 8-9. Chromium (VI) is then removed by contact with a strong base ion exchanger (hydroxide form) and the resin is regenerated with a sodium hydroxide solution. The spent regeneration solution, which contains sodium chromate and sodium hydroxide, is neutralized with sulfuric acid and then treated with a cation exchanger to produce chromic acid for recycle to the plating baths [Arthur D. Little, Inc., 1976]. Hall, et al. (1979) quote a study in which a chromic acid recovery efficiency of 99.5 percent has been demonstrated. Reduction in chromic acid purchases of 76-90 percent was also reported. Nitric acid can be recovered in a similar manner.

Recovery of valuable metals such as chromium, nickel, silver, and gold by ion exchange is more economically attractive than conventional waste

treatment processes [Fisher and McGarvey, 1967]. A
survey conducted by Plating and Surface Finishing
Journal [Anonymous, 1979] showed a favorable pay
back for recovering nickel and gold with ion
exchange equipment.

Ion exchange is more attractive economically
for the removal of heavy metals than the lighter
ones when based on a weight basis, since the capacity of an ion exchanger is based on equivalent
weight. For example, the weight of silver is 107
and of sodium 23, so that an exchanger can pick up
four times the weight of silver compared to the
weight of sodium it can hold [Gold and Calmon,
1980].

Another large scale application of the ion
exchange process is the removal of aluminum from
strong phosphoric acid/nitric acid solution used
for cleaning ("bright dipping") aluminum metal
[Skovronek and Stinson, 1977]. The contaminated
phosphoric acid is diluted with rinse tank water to
give a 40 percent phosphoric acid solution which is
then subjected to acid exchange to remove the aluminum. The purified phosphoric/nitric acid mixture
is then evaporated to produce concentrated acid for
recycle to the process. The resin is regenerated
with sulfuric acid to yield aluminum sulfate.

The ion exchange technique can be improved in
terms of its utility and economics by using some
ingenuity in design. For example, an ion exchange
system, called "Reciprocating Flow Ion Exchange"
has been perfected by Eco-Tec Limited for recovery
of waste metals from plating operations [Brown,
1975].

There are a number of new developments in the
areas of metal pickling waste treatment, cyanide
removal, and novel ion exchange resins. A number
of new ion exchange resins are being tested on a
laboratory and pilot scale [Gold and Calmon, 1980].
Many of these resins are of the macroreticular type
and less subject to fouling and loss of capacity
than the older materials.

Limiting Technology

The ion exchange process is a well-established
process used in various industrial operations.
However, the major limiting factor in applying this

technique to new situations seems to be the presence of materials or conditions which may clog, attack, or foul the resins. For example, high concentrations of oxidizing agents such as chromic acid can damage the resins. Active research is currently being conducted to evaluate the use of new ion exchange materials, which would not be affected by the presence of such materials.

There seems to be a need for the development of highly basic anion exchangers which are stable at elevated temperatures when in the base form. Highly basic anion exchangers tend to foul with some surface waters and in solutions containing high molecular anionic species. Therefore, anionic exchangers less subject to fouling are needed, which would increase applications in solutions containing foulants [Gold and Calmon, 1980].

More research is warranted to develop continuous ion exchange systems requiring less technical supervision and more controllability than those which exist.

In some cases the regenerant or eluting solution is not economically worth recovering or reusing. It can, however, become a pollutant if discarded. Therefore, more research is needed on full utilization of regenerants, higher regenerant efficiency, and reuse of regenerant effluents containing unused regenerant.

Recycle, Recovery, and Reuse Applications

The ion exchange process has been used for several years by different industries. Ion exchange is currently used for both general and selective removal of primarily inorganic ion species. Applications of ion exchange process in the waste treatment area include: treatment of a wide variety of dilute wastewaters from electroplating and other metal finishing operations, recovery of materials from the fertilizer manufacturing industry and hydrometallurgical processes, removal of cyanide from mixed waste streams, and recovery of chromium from cooling tower blowdown.

Reverse Osmosis

Osmosis is defined as the spontaneous transport of a solvent from a dilute solution to a concentrated solution across an ideal semipermeable membrane, which impedes passage of solute but allows solvent flow [Weber, 1972]. At a certain pressure exerted on the solution side of the membrane, called osmotic pressure, equilibrium is reached and the amount of solvent which passes in each direction is equal. If the pressure is increased above the osmotic pressure on the solution side of the membrane, the flow reverses. Pure solvent will then pass from the concentrated solution to the dilute solution. This is the underlying principle of the reverse osmosis.

The solvent flux (frequently expressed in gallons/ft -day) is given by [Arthur D. Little, Inc., 1976]:

$$J = K \ (\Delta P)\pi \qquad\qquad (6.5)$$

In which ΔP is the applied pressure, π the osmotic pressure, and K is a constant for the membrane-solvent system. As can be seen from this equation, the product-water flux rate decreases with increasing salinity (increasing osmotic pressure) of the feed solution.

The basic components of a reverse osmosis unit are the membrane, a membrane support structure, a containing vessel, and a high-pressure pump. Cellulose acetate and nylon are more common among the membrane materials used. The chemical nature of the membrane material is important because it affects the transport of solvent and rejection of solute. The membranes are susceptible to chemical attack and fouling, so pretreatment of certain feedwaters may be necessary. Usually, as a part of pretreatment, iron and manganese salts are removed to decrease fouling potential and the pH is adjusted to a range of 4.0 to 7.5 to inhibit scale formation.

Process Configurations

Based on the membrane support structure, reverse osmosis units may be classified under four categories: spiral-wound, tubular, plate and frame, and hollow fiber. The tubular configuration is recommended for use with domestic wastewater effluents while reverse osmosis system using a multitude of hollow nylon or polyamide fibers, have shown considerable utility on commercial waste

streams. Reverse osmosis units can be arranged
either in parallel to provide adequate hydraulic
capacity or in series to effect the defined degree
of demineralization.

Reverse Osmosis: State-of-the-Art

Research in the area of reverse osmosis has
recently focused upon the development of improved
families of membranes with increased flux, greater
resistance to degradation in various liquid envi-
ronments and improved rejection characteristics.
Progress is also being made in recent years in
improving mechanical designs for the modules, which
can work efficiently with thin membranes at operating pressures of up to 800 psig [Leitner, 1973].

The effect of pH and temperature or membrane
permeability was studied by Beder and Gillespie
(1970). They reported that the flux rate was
slightly higher under alkaline conditions. Their
results also showed that a 5°C deviation from 25°C
produced a 15 percent change in the membrane permeability constant, K. Similar results were re-
poted by Wiley et al. (1967).

Kojima and Tatsumi (1977), who studied the
reverse osmosis treatment of wastewater from a
chemical plant, reported that th performance of the
membrane deteriorated gradually due to fouling and
compaction of the membrane. It was found that the
presence of organic matter, silica, calcium, and
iron in the feed was responsible for the damage of
the membrane. Pretreatment of the feed to reverse
osmosis units is necesssary for almost all industrial wastewaters. Due to the sensitivity of the
membranes to the high concentrations of suspended
solids in wastewaters, diatomaceous filters followed by a sand filter [Warnke, et al. 1976], or
coagulation and sedimentation [Kojima and Tatsumi,
1977], have been used for pretreatment purposes.
Removal of high concentrations of suspended solids
in the feed before reverse osmosis is necessary in
order to prevent clogging of the membrane and subsequent decrease in flux rates. Fouling of the
membrane due to the presence of certain chemicals
can be overcome by precipitating and filtering such
materials by chemical treatment [Zabban and
Helwick, 1980].

Statement of Limiting Technology

Reverse osmosis is a relatively new process and the development of this process has accelerated within the past few years, with a substantial increase in the number of commercial installations. There are, however, certain practical limitations to the use of reverse osmosis for waste recovery and reuse.

There seems to be a limitation on the pressures used in the reverse osmosis process. At high pressures, the membranes are subjected to compaction, which is accompanied by a decrease in product flux. Most units are limited to moderate temperature, slightly acidic pH, and influent which can be prefiltered effectively to prevent fouling of membranes [Hall, et al. 1979].

The membranes available in the market at the present time are not sufficiently resistant to a wide range of chemicals, such as are often found in waste streams. These membranes do not seem to withstand extreme pH, temperature, and pressure. Research is needed to develop new membranes which will broaden the application of reverse osmosis.

Due to the limited applicability of the membranes, most of the industrial waste streams need some type of pretreatment prior to entering the reverse osmosis unit. Such pretreatment would add to the overall cost of the treatment system and might pose a limitation to the use of reverse osmosis.

The concentrate from the reverse osmosis operation, in some cases, must be further concentrated in an evaporator or subjected to some other treatment for a complete recovery of the pollutants. The limitations of reverse osmosis systems used in various industrial waste treatment operations are summarized in Table 6.1.

Recycle, Recovery, and Reuse Applications

Reverse osmosis has gained importance as an industrial unit operation only in recent years. Initially, it was used for production of potable water. However, it is currently used in a broad spectrum of industrial opeations. Recent research indicates wastewater reclamation by reverse osmosis

Table 6.1   Evaluation of RO for Systems Tested
[Skovronek and Stinson, 1977].

Attractive Systems	Limitations
Watts-Type Nickel	Boric acid selectively permeates membranes
Nickel Sulfamate	Boric acid selectivelypermeates membranes
Copper Pyrophosphate	Possibledecomposition ofpyrophosphate
Nickel Fluoborate	Boric acidselectivelypermeates membranes
Zinc Chloride	Need evaporation to close loop
Copper Cyanide	Need low-pH bath for current membranes:
Zinc Cyanide	Need low-pH bath for current membranes: need evaporation to close loop
Cadmium Cyanide	Need low-pH bath for current membranes: need evaporation to close loop
Unattractive Systems	Limitations
Chromic Acid	Attacks and destroys all membranes unless neutralized
Very-high pH Cyanide Baths	Attacks and destroys all membranes commercially available: newer membranes under development show promise for treating high-pH cyanidebaths

offers great promise for substantial reductions in cost as well as marked improvements of pollutant removal efficiency.

Reverse osmosis has been used for various industrial operations and its use is expected to expand as its utility is demonstrated and its technology becomes more familiar to potential users. Some of the applications of reverse osmosis in waste treatment area include: plating wastes, paper mill effluents, laundry wastewaters, food processing wastes, acid mine drainage waters, and petrochemical wastewaters.

REFERENCES

Anonymous, "Recovery Pays." Plating and Surface Finishing, 662, 4548, 1979.

Arthur D. Little, Inc., Physical, Chemical and Biological Treatment Techniques in Industrial Wastes, Vol. II. National Technical Information Service: PB275 287, 1976.

Beder, H. and Gillespie, W.J. "Removal of Solutes from Mill Effluents by Reverse Osmosis." Tappi, 53, 5, 883-887, 1970.

Berkonitz, J.B. et al. Physical, Chemical, and Biological Treatment Techniques for Industrial Wastes. NTIS No. PB275054, Arthur D. Little, Inc., Cambridge, Mass., 1976.

Brown, C. "Effective Nickel Recovery Will Prove Profitable." Plant Management and Engineering, 34, 8, 2729, 39, 1975.

Chian, E.S., and Fang, H.H. "Physicochemical Criteria Removal." AIChE Symp. Series, No. 166, Vol. 73, p. 152, 1977.

Cruver, J.E. "Reverse Osmosis for Water Reuse." AIChE Conf. on Complete Watereuse, p. 619, 1973.

Fels, A. "Permeation and Separation Behavior of Binary Organic Mixtures in Polyethylene." AIChE Symp. Series, No. 120, Vol. 68, p. 49, 1972.

Fenton-May, R.I., Hill, C.G., Amundson, C.H., and
    Auclair, P.D., "The Use of Pressure Driven
    Membrane Processes in the Dairy Industry."
    AIChE Symp. Series, No. 120, Vol. 68, p. 31,
    1972.

Gold, H. and Calmon, C. "Ion Exchange: Present
    Status, Needs, and Trends. In: Recent Advances
    in Separation Techniques." AIChE Symp.
    Series, No. 192, Vol. 76, p. 60-67, 1980.

Hall, E.P., Lizdas, D.J., and Auerbach, E.E.
    "Recovery Techniques in Electroplating."
    Plating and Surface Finishing, 66, 2, 49-53,
    1979.

Kojima, Y. and Tatsumi, M. "Operation of Reverse
    Osmosis Process for Industrial Waste Water
    Reclamation." Desalination, 23, 8795, 1977.

Leitner, G.F. "Reverse Osmosis for Water Recovery
    and Reuse." Chemical Engineering Progress,
    69, 6, 83-85, 1973.

Mir, L., Eykamp, W. and Goldsmith, R.L. "Current
    and Developing Applications for
    Ultrafiltration." Ind. Water Eng. 14(3):14,
    May/June, 1977.

Nelson, R.F. "Ultrafiltration for Polyglycol
    Removal." AIChE Conf. on Complete Watereuse,
    p. 926, 1973.

Ohya, H., Akimoto, N., and Negishi, Y. "Reverse
    Osmosis Characteristics of Cellulose Acetate
    Butyrate Membranes." J. Applied Polymer
    Science, 24(3):663, 1 Aug., 1979.

Porter, M.C. "Ultrafiltration of Colloidal
    Suspensions." AIChE Symp. Series, No. 120,
    Vol. 68, p. 21, 1972.

Raman, R. and Karlson, E.L. "Reclamation of
    Chromic Acid Using Continuous Ion Exchange."
    Plating and Surface Finishing, 64, 6, 40 and
    42, 1977.

Skovronek, H.S. and Stinson, M.K.   "Advanced
    Treatment Approaches for Metal Finishing
    Wastewaters: Part 2."   Plating and Surface
    Finishing, 64, 11, 24-31, 1977.

Sundstrom, D.W. and Klei, H.E.   Wastewater
    Treatment.   Prentice-Hall, Inc., Englewood
    Cliffs, New Jersey, 1979.

Warnke, J.E., Thomas, K.G., and Creason, S.C.
    "Reclaiming Plating Wastewater by Reverse
    Osmosis."   Proc. 31st Industrial Waste
    Conference, Ann Arbor Science Publishers,
    Inc., Ann Arbor, Michigan, pp. 525-530, 1976.

Weber, W.J., Jr. Physicochemical Processes For
    Water Quality Control.   Wiley-Interscience,
    New York, New York, 1972.

Wiley, A.J., Ammerlain, A.C.F., and Dubey, G.A.
    Tappi.   50, 9, 455, 1967.

Zabban, W. and Helwick, R.   "Cyanide Waste
    Treatment Technology-The Old, The New, and The
    Practical."   Plating and Surface Finishing,
    67, 8, 56-59, 1980.

SUPPLEMENTAL REFERENCES

Bailey, P.A.   "Ultrafiltration - The Current State
    of the Art."   Filtration and Separation.   Vol.
    14(3):213, May/June, 1977.

Brandon, C.A. and Samfield, M.   "Application of
    High-Temperature Hyperfiltration to Unit
    Textile Processes for Direct Recycle."
    Desalination. 24(1/2/3):97, Jan. 1978.

Goldsmith, R.L., de Filippi, R.P., and Hossain, S.
    "New Membrane Process Applications." AIChE
    Symp. Series, No. 120, Vol. 68, p. 7, 1972.

Gross, M.C., Markind, J., and Stana, R.R.
    "Membrane Experience in Food Processing."
    AIChE Symp. Series, No. 129, Vol. 69, p. 81,
    1973.

Masuda, H., Kamyawi, C., Hata, K., Yokota, K., Sakai, T. and Soto, M. "Concentration of Acetic Acid in Sulfite Pulp Evaporation Drain by Reverse Osmosis." Desalination 25(1):89, Mar. 1978.

Murkes, "Some Viewpoints on the Industrial Application of Membrane Technology." Desalination 24(1/2/3):225, Jan. 1978.

Spatz, D.D. "Reclamation and Reuse of Waste Products from Food Processing by Membrane Processes." AIChE Symp. Series, No. 129, Vol. 69, p. 89, 1973.

INTRODUCTION

In this chapter, four processes will be examined which rely on one (or several) constituents of a wastestream to undergo a phase change in order to separate it (and in many cases these constituents are recycled and reused) from the main fluid stream. The processes to be examined include:

1. Condensation
2. Distillation
3. Evaporation
4. Refrigeration

Processes of heat transfer accompanied by phase changes are more complex than simple heat exchange between fluids. A phase change involves the addition or subtraction of considerable quantities of thermal energy at constant or nearly constant temperature.

Condensation, evaporation and refrigeration are processes used primarily to separate constituents from a wastestream. Once separated, these constituents can be further purified and either recycled back into the process or sold. Evaporation, on the other hand, is basically used to concentrate a non-volatile solute by vaporizing the volatile solvent. The solvent can then be condensed, purified and recycled into the process for further use. The concentrate (non-volatile solute and remaining solvent) can also be recycled, or disposed of, at this point.

For each of the four listed processes, a brief process description is provided, plus a review of the current recycle, recovery, and reuse applications, and a statement on the processes limiting technology.

CONDENSATION

In a two component vapor containing one condensable and one noncondensable component, condensation will occur when the partial pressure of the condensable component equals that component's vapor pressure. This may be effected in one of two ways: The system pressure may be increased at a given temperature until the partial pressure of the condensable component equals its vapor pressure. Alternately, and far more commonly, the pressure remains fixed and the temperature of the mixture is reduced to the point where the vapor pressure of the condensable component equals its partial pressure. At further reduction in temperature, condensation continues so that the partial pressure is always equal to the vapor pressure [Anonymous, 1978].

Condensers may employ contact or non-contact methods for cooling the vapor. Contact condensers usually spray an ambient-temperature or slightly chilled water or other liquid directly into the gas stream in order to condense the vapor. The contact condenser may also act as an absorptive system, scrubbing vapors which might not be condensed, but which are soluble in the liquid. Because the temperature approach between the liquid and the vapor is very small, condenser efficiency is high. However, large volumes of liquids are required.

Direct contact condensers are seldom used for the removal of organic solvent vapors because the condensate will contain an organic-water mixture which must be separated or treated before disposal. They are effective in cases where it is necessary to remove heat from hot gas streams without concern for recovery of organics. Spray towers, high velocity jets, and barometric condensers are among the equipment used for contact condensation. In practice, crude empirical correlations are usually used in designing and predicting performance of a contact unit [Theodore, and Buonicore, 1975].

Surface condensers are non-contact units, which may also be referred to as heat exchangers. A common surface condenser is the shell-and tube heat exchanger in which the coolant flows through the tubes and vapor condenses on the outer tube surface. The film of condensed vapor which develops drains away to storage or disposal. In this way, the coolant contacts neither the vapor nor the condensate. Surface condensers may also be air-cooled. These air-cooled units usually have extended surface fins; vapor condenses within the finned tubes [Theodore and Buonicore, 1978].

Although contact condensers are generally less expensive, easier to maintain, more flexible, and more efficient in removing organic vapors than surface condensers, surface condensers are more attractive with respect to recovery of marketable condensate and yielding a minimal waste disposal problem [Anonymous, 1978]. Furthermore, water used as a surface condenser coolant may be reused, and surface condensers produce 10 to 20 times less condensate than do contact condensers [Theodore and Buonicore, 1978]. Because high removal efficiencies are not obtainable with low condensable vapor concentrations, condensers are typically used when the vapor concentration in the stream to be treated is high, and are often located upstream of after burners, carbon beds, or adsorbers as a pretreatment measure.

Applications

A shell and tube heat exchanger was installed on a neoprene monomer in somerization tower to treat a total waste gas flow of 331 lb/hr. The hydrocarbon waste gas flow was 159 lb/hr. The exchanger, cooled by -2°F brine solution, condensed 81 percent of the contained hydrocarbon with an energy requirement of 22,000 Btu/hr and the recovered hydrocarbons were returned to the process for utilization [Pruessner and Broz, 1977].

In a process in which fuels are produced from coal, volatile products released from the coal in fluidized bed reactors pass to a product recovery system for recovering the oil and cooling the gases. The coal-oil vapors are directly contacted with a water-rich stream for condensation. The oil-phase is dehydrated and filtered to remove solids before being pumped up to pressure and mixed with hydrogen for hydrotreating in a fixed-bed catalytic reactor. Hydrotreating removes sulfur, nitrogen, and oxygen to produce a synthetic crude oil [Jones, 1974].

Power plant turbine exhaust steam has been mixed with cooling water in a direct contact

condenser maintained under a vacuum. The temperature
of water is less than the boiling temperature under
the vacuum conditions in the condenser. The mixture of cooling water and condensed turbine exhaust
is divided so that one portion is sent to the
boiler, suitable for use in generating steam to
drive the turbine, while the other portion is sent
to the condenser for use as cooling water after it
has passed through an air cooled heat exchanger.
Water to be used for cooling is placed in the heat
exchanger with liquid gaseous fuel so that the fuel
is vaporized at the expense of heat lost from the
water [Anonymous, 1974].

In work on liquid fuel synthesis using nuclear
power in a mobile energy system, Steinberg and
Beller proposed to extract CO_2 from the atmosphere
by compressing air, condensing the water from it,
drying the resultant air with a molecular sieve,
and extracting CO_2 by another molecular sieve
[Steinberg and Beller, 1964]. Williams and
Campagne (1971), however, found compressor costs
for such an operation to be prohibitive.

At a neoprene polymer plant where neoprene
latex is manufactured in batch reactors, each
charging and emptying of a polymerization reactor
causes displacement of some waste gas. Collected
waste gas is discharged to a direct contact cooler.
A problem which precluded the use of a shell and
tube exchanger was the regular carryover of latex
to the collection system, anticipated to produce a
high rate of fouling. The waste gas flow of 275
lb/hr., containing about 125 lb/hr hydrocarbon
contamination, is cooled with chilled water, resulting in the condensation of 43 percent of the
hydrocarbon. Exit gas from the direct contact
cooler is treated in an absorption system.

The absorption system consists of a five-stage
oil absorption tower. Overall efficiency of the
contact cooler/absorption system exceeds 98 percent
hydrocarbon removal. The recovered hydrocarbon is
stripped from the oil and returned to the manufacturing process for further use [Pruessner and Broz,
1977].

A solvent recovery system features an inert
nitrogen atmosphere in an oven/dryer process. In
this process, resin curing is accomplished by evaporating the organic solvent in which the resin is

dissolved. This eliminates the requirement for
oven ventililation with atmospheric air, normally
required to dilute solvent concentrations to below-
explosive levels. By using the inert nitrogen atmosphere, solvent vapor can be safely concentrated to
well above traditional oven levels. At these
levels, the solvent is recoverable by condensation.
According to the manufacturer, solvent recovery is
about 99 percent, and overall fuel requirements are
reduced by 40 percent [Anonymous, 1980].

Limiting Technology

     High removal efficiencies are not obtainable
with low concentrations of condensable vapors.
This limits the application of the condensation
process to streams having a high vapor concentration or to pretreatment of streams fed to after-
burners, carbon beds or other absorbers.

DISTILLATION

     Distillation is a unit operation employed by
industry for the separation, segregation, or purification of liquid organic product streams, some of
which may contain aqueous fractions. The term
distillation is properly applied only to those
operations in which vaporization of a liquid mixture yields a vapor phase containing more than one
component. This distinguishes the process from
that of evaporation.

     As a unit operation, distillation has been
successfully used either singly or in combination
with such operations as direct condensation, adsorption, and absorption for the recovery of organic solvents. With regulations which are becoming increasingly stringent for air pollution
control, liquid effluents, land site disposal, and
the rising cost of organic chemicals, distillation
should become more competitive with other methods
of organic liquid recovery and disposal.

     The practical limitations of the process are
economic in nature; both operational and equipment
costs are high. The difficulty in separating the
contents in the liquid to be distilled defines the
economic and energy requirements of the process
[Berkowitz, et al. 1976].

Theoretically, distillation can generate products of 100 percent purity. Physical parameter restrictions such as entrainment effects limit the degree of attainable purity, however.

In terms of the physical separation of liquid components, there are no limitations in feed composition or in reaching the desired composition in any of the product streams. To avoid plugging of equipment or costly maintenance problems, it is preferred that materials to be distilled do not contain appreciable quantities of solids or non-volatile materials, and that the feed does not have a tendency to polymerize. If it is known that feeds of a "dirty" nature must be handled, pre-treatment steps employing filtration, thin-film evaporation, scrubbers, electrostatic precipitators, or cyclones may be taken. When pretreatment is not possible, specially designed equipment may be required.

Types of Distillation Processes

The most common types of distillation are those of batch and continuous fractional distillation. Certain feed streams require more specialized processing, however.

When it is necessary to distill an azeotrope, pressure or vacuum may be applied to shift the azeotropic composition. More often, an additive is introduced to the mixture to form a new boiling point azeotrope with one of the original constituents. The volatility of the new azeotrope is such that it may be easily separated from other original constituents.

Binary mixtures which are difficult or impossible to separate may also be distilled by extractive distillation. In this process, a solvent is added to the mixture which alters the relative volatility of the original constituents, thus permitting separation. The added solvent is of low volatility and not appreciably vaporized.

In the case of heat sensitive feed stream, molecular distillation may be used. This process is conducted at absolute pressure on the order of 0.003 mm of mercury [Berkowitz, et al. 1976].

Limiting Technology

The fact that the distillation process is energy intensive tends to limit its application, but with increases in by-product recovery credits, it may become competitive with more commonly utilized processes. Organic peroxides, pyrophoric organics, and inorganic wastes cannot generally be treated by distillation.

Recycle, Recovery, and Reuse Applications

Typical industrial wastes which can be handled by distillation include:

- Plating wastes containing an organic component - usually the solvents are evaporated and the organic vapors distilled.

- Organic effluents from printed circuit boards are adsorbed on activated carbon. Regeneration of the activated carbon gives a liquid which is distillable for recovery of the organic component.

- Phenol recovery from aqueous solutions is a major waste treatment problem. The recovery process uses a polymeric adsorber. The adsorber is regenerated using a vaporized organic solvent. A complex distillation system is used to recover both the regeneration solvent and the phenol.

- Methylene chloride which contains contaminants is a disposal problem, but it can be salvaged for industrial application by distilling.

- Methylene chloride can be recovered from poly urethane waste by distillation.

- The separation of ethylbenzene from styrene and recovery of both by distillation is possible.

- Waste solvents recovery by distillation is one of the principal applications of this technology.

EVAPORATION

Evaporation is the vaporization of a liquid from a solution or a slurry for separation of liquid from a dissolved or suspended solid or liquid. The basic principle underlying evaporation is to concentrate a solution consisting of a non-volatile solute and volatile solvent [Arthur D. Little, Inc., 1976]. This is usually achieved by condensation of steam on metal tubes, which have the material to be evaporated flowing inside them. The solvent, which is in a vaporized form after the evaporation process, may be discharged as an exhaust or can be condensed, purified, if necessary, and reused. Similarly, the residue from the evaporation process, which is called the concentrate, can be disposed of or the useful materials in it may be reused. A typical example where both the condensate and the concentrate can be recycled back to the original process where these materials are used is the electroplating industry.

Process Configurations

Basically, there are two types of evaporation systems, atmospheric evaporation and vacuum evaporation. Atmospheric evaporation can be accomplished simply by boiling the liquid. Evaporation can be achieved at lower temperatures by spraying the heated liquid on a surface and blowing air over the same surface.

In vacuum evaporators, the boiling temperature is reduced by lowering the evaporating pressure. The water vapor is condensed and the non-condensible gases are removed by a vacuum pump. Vacuum evaporation may be either single or multiple effect.

Single effect evaporators are used where the required capacity is small, steam is cheap, the vapors on the liquids are so corrosive that very expensive materials of construction are required, or when the vapor is so contaminated that it cannot be reused [Arthur D. Little, Inc., 1976]. In a multiple-effect evaporator, steam from an outside source is condensed in the heating element of the first effect. The vapor produced in the first effect is used as the heating medium of the second effect, which is operating at a lower pressure than the first effect. The vapor from the second effect

is used as the heating medium for the third, and so
on. Each consecutive effect operates at a lower
pressure than the preceding effect.

Several types of evaporators are used for the
separation and recovery of different organic and
inorganic compounds. In process industries, such
as metal and plastics finishing industries and
particularly the electroplating industry, closed-
loop recycling is achieved by using evaporation as
a recovery process. In a closed-loop system, the
condensate and the concentrate are recycled back to
the rinsing and plating operations respectively,
thereby forming a closed-loop treatment.

Evaporation: State-of-the-Art

A survey done by Plating and Surface Finishing
Journal [Anonymous, 1979] concluded that a favorable payback for evaporative equipment is achieved
for the recovery of chromium, nickel, and lead-tin
alloy plating chemicals but not for zinc cyanide
solutions. It was felt that operating costs for
recovery equipment exceeded the saving in zinc
plating chemicals. However, recovery may be justi-
fied because it eliminates costly cyanide destruction, zinc precipitation and solids separation and
sludge disposal. There does not seem to be much
information available in this area and more research is needed to study the application and cost
analysis of evaporation systems for recovering zinc
plating chemicals.

Recovery of industrial pollutants by evaporative techniques becomes more cost-effective when
the pollutants in the wastes are in high concentrations and the flow rates are low. In order to
achieve this, multistage countercurrent rinse tanks
are suggested in the case of the electroplating
industry.

As mentioned earlier, one of the major limitations of evaporation techniques is the cost of
heating from steam generators. This may be substantially reduced by utilizing waste heat from the
plating baths of the electrochemical industry
[Cheremisinoff, et al. 1977; Hall, et al. 1979;
Skornonek and Stinson, 1977]. More research is
warranted in this area to determine the feasibility
and economics of utilization of waste heat.

Another way to economize on energy consumption is to use the multiple-effect evaporation technique, in which the vapor produced from one effect is used as the heating medium for the next one. In fact, one electroplating company abandoned its use of a single-effect evaporator for nickel recovery because operational costs were greater than the value of the recovered nickel, and calcium ions and other impurities resulted in fouling. This company now uses three counter-flow rinses to concentrate and recover nickel salts from its rinse water waste and periodically dumps the fourth final rinse to control impurities [Anonymous, 1979].

In the case of cyanide copper plating operations, certain types of evaporation techniques (for example, climbing film evaporators), may not be used since the materials of construction in use are appropriate only for acid pH. Different types of materials are needed in such cases [Hall, et al. 1979].

Process Modifications

Ideally, evaporation results in a concentrate and a purified condensate. But, in a practical situation carry over of some impurities in the condensate may occur. Furthermore, the concentrate may also contain contaminants such as organic brighteners and anti-forming agents. An activated carbon bed may be necessary to remove such impurities before the condensate is recycled back [Hall, et al., 1979]. In order to recover copper from copper cyanide plating operations, the concentrate may be treated with peroxide, caustic potash, and activated carbon [Hall, et al. 1979].

In the case of nickel plating operations, [Atimion, 1980] reported that when the nickel solution is sufficiently concentrated from evaporation, it is pumped from the evaporator to a stainless steel tank for carbon treatment at pH 1.5 to 1.7. The solution is then filtered into a clean tank and nickel carbonate is added to the solution to raise the pH to 5.0 to 5.5. Then, required doses of potassium permanganate and activated carbon are added to remove sulfur. After settling for eight hours, the treated solution is filtered twice and recycled to the plating bath.

Pretreatment of the wastewater is sometimes necessary to make the evaporation process applicable to a given situation. For example, most of the companies using evaporators to recover chromium solution have installed cation exchangers to control the concentration of trivalent chromium and other metallic impurities that tend to reduce cathode efficiency [Anonymous, 1979; Hall, et al. 1979; Kolesar, 1972; Cheremisinoff, et al. 1977]. In the case of cyanide plating, purification is accomplished by precipitaton of carbonate, which is best done at the point of maximum concentration [Cheremisinoff, et al. 1977].

Limiting Technology

There do not seem to be any fundamental limitations on the applicability of evaporation process. However, energy consumption appears to be the major obstacle. There are certain practical limitations in the application of evaporators. In addition to heat transfer characteristics and economic energy utilization, the other variables that limit the practical application of evaporation process are crystal formation, salting, scaling, corrosion entrainment, and foaming [Arthur D. Little, Inc., 1976].

In order to prevent these problems in evaporators, studies should be conducted to determine the proper type of evaporation for a given industrial operation. Information in this area seems to be scarce.

It appears that most of the work on application of evaporation techniques used for the recovery of industrial pollutants has been done in the elctroplating industry. Evaporation systems are also used by the paper industry to recover the chemicals from the Kraft pulping process [Arthur D. Little, Inc., 1976].

REFRIGERATION

The process of cooling, or heat withdrawal, may be accomplished by the solution, melting or evaporation of a substance, or by the extension of a gas. The term refrigeration refers particularly to cooling below atmospheric temperature. Machines which produce cooling may be classified into compression and adsorption refrigeration machines,

depending on the mode of recovery and circulation
of material.

Compression refrigeration machines evaporate
low-boiling liquids and condense their vapors or
gases. Mechanical energy is used to effect compression of the refrigerant. The refrigerant is
then sent from the condensor to the evaporator
where it withdraws heat by evaporation and is
recycled for compression. Electric energy may also
be used to compress the refrigerant, and turbo-
compressors driven by electric motors or steam
turbines may be used instead of piston compressors.

Adsorption refrigerating machines utilize heat
to produce cold, and operate economically when waste
heat is available. They are especially useful for
low temperature evaporation pressures. Refrigerant
leaving the evaporator travels to an adsorber where
it is bound by the aid of a liquid pump. After
passing through a heat exchange, it travels to a
stripping tower, where the refrigerant is stripped
from the water by the addition of heat. The gas is
then liquefied in the condensor. The mechanical
compressor used in compression refrigerating
machines is substituted by a thermal compressor in
adsorption refrigerating machines [Perry, 1963].

In practice, vapors undergoing refrigeration
are condensed by either contacting a cold surface
or by contact with the coolant. When a vapor is
refrigerated for the purpose of emission control or
product recovery, the unit in which heat transfer
takes place between the refrigerant and the vapor
is referred to as the condensor. In a surface
condensor, the coolant does not contact the vapors
or condensate. Coolant, vapors and condensate are
intimately mixed in contact condensors.

The choice of condensor will be influenced by
a) the presence of moisture with the vapor, b) the
operating temperature level, and c) whether or not
the condensed vapors are to be reused as product
liquid. Cooling an air-vapor stream condenses the
moisture to liquid at a temperature above 32° F.
This may lead to a gradual build up of frost which
can be removed by scrapers or by periodically
raising the surface temperature above 32° F. A
contact condensor does not lend itself to frost
removal, since any solid which is formed circulates
with the contact and condensate, impairing the

performance of pumps and valves [Honegger, 1979].

Limiting Technology

In the case of hydrocarbon recovery, removal efficiency by refrigeration depends on the hydro- carbon concentration of inlet vapors. The degree of product recovery by refrigeration is also depen- dent on the refrigeration temperature level. The more volatile the product, the lower the tempera- ture required for effective recovery. When mois- ture is entrailed in the vapor, special equipment may be necessary to accommmodate the ice and frost which build up, impairing process performance.

Recycle, Recovery, and Reuse Applications

Refrigeration is one of several competing methods for recovering emissions from bulk liquid transfer and storage operations, and has been pro- moted for vapor recovery at gasoline loading racks. Equipment and operating costs are relatively high for this application because ultra-low temperatures are required for effective recovery [Honegger, 1979].

At some gasoline terminals, vapors are com- pressed and then refrigerated to obtain condensa- tion. Other installations omit compression and refrigerate the vapors to temperature approaching $73°$ C ($100°$F). Removal efficiencies depend on the hydrocarbon concentration of inlet vapors. In the case of saturated hydrocarbons, removals of greater than 96 percent are possible. Similar systems have been proposed for marine petroleum terminals [Anonymous, 1978].

The degree of product recovery is also depen- dent on the refrigeration temperature level. The more volatile the product, the lower the tempera- ture required for effective recovery. Since many hydrocarbon vapors require temperatures below the point at which water freezes, special equipment may be necessary to accommodate the ice and frost which form due to moisture entrained in the air [Honegger, 1979].

REFERENCES

Anonymous (1974) Condensation of Stream Turbine Exhausts British Pat. 1,361,825.

Anonymous (1978) Control Techniques for Volatile Organic Emissions from Stationary Sources. U.S. EPA, Office of Air and Waste Management, EPA 450/2-78-022.

Anonymous, (1979) "Recovery Pays." Plating and Surface Finishing, 66, 2, 45-48.

Arthur D. Little, Inc., (1976) "Physical, Chemical and Biological Treatment Techniques in Industrial Wastes, Vol. II." National Technical Information Service: PB-275, 287.

Atimion, L., (1980) "A Program of Conservation, Pollution Abatement." Plating and Surface Finishing, 67, 3, 18-20.

Berkowitz, J.B., et al., (Nov., 1976) Physical, Chemical, and Biological Treatment Technique for Industrial Wastes. NTIS #PB-275 054.

Cheremisinoff, P.N., A.J. Perna, and J. Ciancia, (1977) "Treating Metal Finishing Wastes, Part 2." Industrial Wastes, 23, 1, 32-34.

Hall, E.P., D.J. Lizdas and E.E. Auerbach, (1979) "Recovery Techniques in Electroplating." Plating and Surface Finishing, 66, 2, 49-53.

Honegger, R.J., (1979). "Refrigeration Methods of Vapor Recovery," Technol. Rept. Gard Inc., A subsidiary of GATX.

Jones, J.F. (1974) "Clean Fuels from Coal for Power Generation". Preprint, Amer. Chem. Soc. (Presented at the Amer. Chem. Soc. Storch Award Symp., Atlantic City, N.J., Sept. 11).

Kolesar, T.J., (1972) "Closed-Loop Recycling of Plating Wastes." Industrial Finishing, 48, 9, 22-25.

Perry's Chemical Eng. Handbook. (1963) McGraw Hill.

Pruessner, R.F. and L.O. Broz (1977) "Hydrocarbon Emission Reduction Systems." J. Chem. Eng. Prog. 73(8):69-73.

Skovronek, H.A., and M.K. Stinson, (1977)
    "Advanced Treatment Approaches for Metal
    Finishing Wastewaters: Part 1." Plating and
    Surface Finishing, 64, 10, 30-38.

Steinberg, M., and Beller, M. (1964) Brookhaven
    National Laboratory Report.

Theodore, L., and A. Buonicore (1975) "Vapor
    Control by Condensation Performance Equations
    and Design Procedures" Proc. APCA 68th Vol. 2
    paper 75-23.2.

Williams, K.R. and N. Van Lookeren Campagne (1971)
    "Synthetic Fuels from Atmospheric Carbon
    Dioxide". Preprint, Shell Intl. Petrol Co.,
    Ltd., London and Shell Intl. Petrol
    Maatshappig N.V., The Hague (Netherlands).

CHAPTER 8

CHEMICAL MODIFICATIONS

INTRODUCTION

This chapter examines several chemical treatment processes currently being used to recover heavy metals from industrial sludges and wastewaters. The processes discussed include:

1. Cementation
2. Precipitation
3. Catalytic Hydrogenation
4. Reduction

For each of the above listed processes, the following areas are discussed:

1. Process Description
2. Recycle, Recovery, and Reuse Applications
3. Statement of Limiting Technology

The "Recycle, Recovery and Reuse Applications" section addresses past and current application of each process in the recovery of heavy metals from industrial wastewaters. The "Limiting Technology" section discusses the advantages, disadvantages and limitations of each process in its application of removing as well as recovering the metals.

CEMENTATION

Cementation involves the recovery of an ionized metal in solution by spontaneous electrochemical reaction to its elemental state through the oxidation of another elemental sacrificial metal. The process can be predicted in terms of electrode potentials. The metal with more positive oxidation potential in the electromotive series will pass into solution displacing a metal with a less positive potential. Since, in electrodeposition, the less noble metal goes into solution, scrap iron or aluminum is often chosen as the reductant because of their low cost.

The cementation process is commonly used to recover copper tailings. Streams bearing copper ions are passed through a series of reaction tanks, containing scrap iron, where the copper is removed as elemental metal and the iron is displaced into solution. The reaction is as follows:

$$Cu^{+2} + Fe^0 \rightleftarrows Cu^0 + Fe^{+2} \qquad (8.1)$$

The cementation process generates electrodeposits without external current. For example, an electropotential sufficiently large to deposit copper is produced when copper ions and iron surface are present in the aqueous solution.

The process can be described as a galvanic corrosion cell. A cathodic copper deposit covers the anode except for sub-microscopic regions where iron is the basic metal, in the manner of a short-circuited galvanic cell. These anodic sites effectively migrate along the surface of the iron; therefore the entire piece of iron can be consumed. The copper deposits can be stripped from the surface of the iron through vigorous stirring or agitation and the resulting copper sludge can be recovered as high purity copper [Patterson and Jancuk, 1977]. Iron is less toxic than copper and can also be readily removed from solution.

If hexavalent chromium is also present along with copper in the wastewater, during the process of cementation, the former may react with either elemental or ferrous iron to yield trivalent chromium [Jester and Taylor, 1973].

$$2\ Cr^{+6} + 3Fe^0 \rightarrow 2Cr^{+3} + 3Fe^{+2} \qquad (8.2)$$

$$Cr^{+6} + 3Fe^{+2} \leftarrow Cr^{+3} + 3Fe^{+3} \qquad (8.3)$$

Limiting Technology

The cementation process has been proven to be effective for the removal and recovery of copper

from concentrated waste streams of small flow. It
has not been exploited for larger flows due to the
need for longer contact time with the metallic
interface.

There seems to be an excess iron consumption
(i.e., more than needed on the basis of stoichio-
metry) associated with the process due to some side
reactions of iron in the wastewater. Disadvantages
of excess iron consumption include wastage of the
metallic iron reactant and production of unnecessary
amounts of iron sludge uponprecipitation treatment.

Thermodynamic limitations and the need for
process optimization seem to prevent the removal of
copper to the low levels required in most effluents. Therefore, the residual copper levels after
cementation treatment would normally require additional treatment. The cementation process for
recovery of metals seems to be in an infant stage
and needs additional study.

Recycle, Recovery, and Reuse Applications

The cementation process is employed to a limited extent by industry today. The hydrometallurgical industry employs the cementation process for
the recovery of such metals as copper on iron,
silver and gold on zinc, lead on iron or zinc,
palladium on copper, thallium on zinc, gold from
gold chloride on cadmium, and gallium on aluminum
[Habashi, 1970].

A few industries have set up wastewater treatment operations for chromium reduction and copper
recovery using the cementation process. The
Anaconda American Brass Company utilizes the cementation process as outlined by their patent for
copper recovery. The copper sludge from their
treatment plant contained 30-40 percent moisture,
with 70-80 percent elemental copper on a dry basis.

The Scovill Manufacturing Company in
Waterbury, Connecticut, adapted the Anaconda approach and designed a process for a simultaneous
chromium reduction and copper removal [Jester and
Taylor, 1973]. The copper sludge from this operation contained more than 95 percent elemental copper by dry weight.

Other known applications of cementation include copper recovery from waste pickle liquor in a
brass mill [Dean, et al., 1972], and the reduction
of cobalt and vanadium with iron metal in the
treatment of waste ammonium, persulfate etching
solutions [Jancuk, 1976].

Industrial application of the cementation
process for copper recovery produce a mud of about
30-40 percent moisture, with dry weight analysis
reported of 95-99 percent [Keyes, 1966], and 70-80
percent [Jester and Taylor, 1973; Case, 1975] pure
copper.

PRECIPITATION

Chemical precipitation is the most common
method for removal of inorganic heavy metals in
industrial waste effluents. Precipitation of a
heavy metal ion occurs when the salt with which it
is in equilibrium reaches its solubility limit, as
defined by its solubility product. The value of
the logarithm of the solubility products of different metal salts are available in the literature
[Bard, 1966; Feitknecht and Schindler, 1963;
Martell and Smith, 1974a, 1974b, 1974c, Sillen and
Martell, 1964, 1974]. These constants may be used
to plot the theoretical solubility diagrams for
each metal, which diagram can be used for determining the pH levels at which each metal is least
soluble.

Theoretical solubility diagrams can be used
only as approximations for determining the pH
levels at which each metal can be precipitated in
the greatest quantity as a given salt, since the
solubility of metals may vary in aqueous solutions
depending upon temperature, ionic strength, and the
presence of anions or other complexing agents in
solution [Butler, 1964; Stumm and Morgan, 1970].

Most of the inorganic heavy metals generally
found in industrial waste effluents are removed by
precipitation as hydroxides. The process of precipitation involves the adjustment of pH of the waste
stream by addition of an acid or an alkali, and
coagulating to effectively remove the resultant
hydroxide through clarification and/or filtration.
Metals may also be removed by precipitation as
sulfides. Metal sulfides are generally more insoluble in water systems than corresponding

hydroxides. However, one of the limitations of sulfide precipitation is that sulfide can form extremely noxious hydrogen sulfide gas.

Chemical precipitation of heavy metals may be accomplished by other batch or continuous treatment systems. Equalization of the waste stream may be necessary for continuous systems if the flow and pH of the waste vary widely with time. The first process step is the adjustment of the pH by addition of acid or alkali to achieve the defined pH level for optimum precipitation. A polymer is usually added to aid coagulation. The waste stream then flows into a sedimentation tank, where the heavy metal precipitate settles out. Precipitated sludge may be recirculated to the precipitation tank in order to provide a seed which will aid agglomeration of the newly formed precipitate.

The chemicals most frequently used for precipitation of metals are lime, sodium hydroxide (caustic soda) and sodium carbonate (soda ash). Lime is preferred because of its relatively lower cost, caustic soda is used in small installations where daily chemical costs are not significant, and soda ash is used in cases where it provides a batch chemical reaction (e.g. cadmium and nickel precipitation) [Lanouette, 1977].

Limiting Technology

Although the solubility of many metallic compounds is extremely low in pure water, such levels often cannot be achieved in industrial effluents through precipitation processes, because of the presence of complexing agents which tie the metals in solution. Furthermore, when two or more metals are found in the same waste stream, the problem becomes even more complex, since the optimum pH for precipitation may be different for each metal salt. The question then becomes whether to treat the waste at a pH which would produce satisfactory, but not optimum results for each of the metal ions present in the wastewater, or to treat it at the optimum pH for one metal ion, remove that metal precipitate, and then treat for the second metal ion and so forth.

Recycle, Recovery, and Reuse Applications

     The precipitation process is used in full-
scale treatment of many industrial wastewaters
containing heavy metals, although not for metals
recovery. Wastewaters from the iron and steel industry [USEPA, 1974], metal finishing industry,
[FWDA, 1968] and inorganic chemical industry [US
EPA, 1975] are some typical examples which receive
precipitation treatment for removal of heavy
metals.

     Cadmium, generated in effluents from pigment
works, textile plants, is usually removed by hydroxide precipitation. The optimum pH for cadmium
hydroxide precipitation is around 11. Such high pH
treatment would require lowering the pH of the
final effluent in order to meet effluent standards.
Since this pH adjustment will redissolve any remaining cadmium, it is essential that the metal
precipitate is removed effectively by filtration
for sedimentation. Due to the high value of cadmium, the precipitated sludge may be sent to a
reprocessor for recovery of cadmium.

     Chromium is found in either the hexavalent or
trivalent form. Hexavalent chromium is present in
the waste streams of plating operations, aluminum
anodizing, and paint and dye operations, while
trivalent chromium is common in effluents from the
photographic, ceramic, and glass industries
[Lanouette, 1977]. Treatment for chromium usually
consists of reduction of hexavalent chromium to its
trivalent form and precipitation of the trivalent
chromium by additon of lime or caustic to increase
the pH to between 7.5 and 8.5, where minimum solubility of chromium hydroxide occurs.

     Lead which is found in waste effluents from
battery manufacture, and printing, painting and
dyeing operations can be precipitated with lime or
caustic soda to form lead hydroxide; with soda ash
to form lead carbonate; or with trisodium phosphate
to form lead phosphate. Lead recovery from the
sludge is possible.

     A standard method of removing mercury is to
adjust the pH to 5 to 6 with H_2SO_4 and then add
sodium sulfide to an excess of 1-3 mg/l. This
forms an insoluble mercury sulfide, from which the
metal can be recovered.

CATALYTIC HYDROGENATION

Catalytic hydrogenation is a useful method for achieving controlled transformation of organic compounds. The reactiono is carried out easily and produces high yield of a product free of contaminating reagents. Satisfactory results can often be obtained over a wide range of conditions of temperature, pressure, and degree of agitation; factors which can affect both activity and selectivity in catalytic hydrogenation.

In any liquid phase hydrogenation, hydrogen moves from the gas phase across a gas-liquid interface and from the liquid phase across a liquid-solid interface to the external surface of the catalyst and then into its porous structure. The net movement of hydrogen to the catalyst results from concentration gradients which develop when hydrogen is consumed by the catalytic reaction. In many cases the reaction rate is limited wholly or partially by the speed with which hydrogen is transported to the catalyst surface. Mass transport is more likely to become a limiting factor as catalytic activity increases.

Due to mass transport limitations on the rate of hydrogenation, hydrogen availability at the catalyst surface may vary from a condition in which the rate of reaction is controlled almost entirely by the rate of chemical reaction to one in which the rate is controlled completely by the rate of hydrogen transport to the active catalyst site.

The process of hydrogenation takes place in a reaction which brings hydrogen, the catalyst, and the substrate into contact in the absence of air. Mosy hydrogenations are carried out in batch-type reactors, although in some cases, especially large-scale processes, continous reactors are used. Reactors may be built for hydrogenation at atmospheric pressure, low pressure, or high pressure. Fixed-bed reactors are useful in the hydrogenation of large volumes of material, and may take the form of a trickle bed, in which a liquid phase and hydrogen flow concurrently downward over a fixed bed of catalyst particles, or a flooded bed, in which hydrogen and the liquid pass concurrently upward.

Hydrogenation catalysts differ widely in activity and selectivity, with these characteristics determined mainly by the major metal component. Metals can be ordered into a hierarchy of activity for hydrogenation of each functional group. Noble metal batch-type catalysts usually contain between 1 and 10 percent metal; fixed-bed catalysts usually contain 0.1-1.0 percent. On a weight-of-metal basis, activity is linear with metal concentration over a limited range. As metal concentration increases, metal becomes piled upon metal, making an increasing percentage unavailable for use. On a weight of metal basis, the more dilute the metal, the more efficient the catalyst, but the amount of catalyst (metal plus support) needed to maintain a constant weight of metal increases directly as the metal concentration decreases. A compromise is made between making the most efficient use of metal and the economic need to minimize the amount of catalyst used. Supported base metal catalysts usually contain much more metal than noble metal catalysts.

Two catalysts used in combination may sometimes give better results than either used separately. Synergism has been explained by the assumption that hydrogenation involves multiple intermediates, some of which may be reduced more easily by one catalyst and some by the other. The second catalyst may also function by its superior ability to remove catalyst inhibitors formed in the reaction.

Catalyst poisons vary from reaction to reaction. Poisons included heavy metal cations, halides, divalent sulfur compounds, carbon monoxide, amines, phosphines, and in some cases, the substrate itself or some product of the reaction. Small amounts of a certain substance may be beneficial to catalyst functioning, while larger amounts will be poisonous. A quantitative measure of catalyst poisons can be made by carrying out the same hydrogenation at different catalyst loadings. If the rate increases more rapidly than the increase in amount of catalyst, the presence of a poison is confirmed.

Small quantities of various substances which favorably affect catalyst life, activity, or selectivity may be termed promoters. The effect a promoter may have also depends on the reaction for which the catalyst is used. There is little theory on which to base the use of promoters, and their successful use usually proceeds from an extension or modification of something already known.

Heterogeneous hydrogenation catalysts may be either supported or unsupported, the former type being further divided into those for use in slurry processes and those for use in fixed bed operations. Catalysts used in slurry processes are usually fine powders, while fixed-bed catalysts are usually in the form of cylinders, spheres, or granules. A good carbon or alumina will be suitable as a catalyst support for the majority of reactions.

Solvents may be used to (1) increase ease of handling and catalyst recovery, (2) moderate exothermic reactions, (3) increase rate and selectivity, and (4) permit hydrogenation of solid material. Most liquids which are stable under hydrogenation conditions and which do not inactivate the catalyst can be used as solvents. Commonly used solvents include acetic acid, methanol, and ethanol.

A problem which may arise during the hydrogenation process is the agglomeration of the catalyst. This will have an adverse effect on the rate and may even cause the reduction to fail. Agglomeration can often be overcome by changing the pH of the medium or by changing the solvent, the solvent-substrate ratio, or the catalyst support.

Lost catalytic activity may, in some cases, be restored by regeneration. Regeneration techniques are basically a variation or combination of oxidations, hydrogenation, steaming, heating, or solvent wash. It is difficult to predict in advance which procedures will work. Eventually, a catalyst can no longer be sufficiently regenerated. Noble metal catalysts can then be returned to a refiner and reclaimed, if volume warrants, for subsequent recovery of the pure metal. Base metals may or may not be reclaimed [Rylander, 1972]. In some industries, spent catalysts are the principal solid waste generated.

Limiting Technology

Catalytic hydrogenation is a useful method for achieving controlled transformation of organic

compounds. Due to mass transport limitations on
the rate of hydrogenation, it may vary from a
condition in which it is almost entirely controlled
by the rate of chemical reaction to one in which it
is controlled completely by the rate of hydrogen
transport to the active catalyst site.

Hydrogenation catalysts differ widely in
activity and selectivity, with these characteristics determined mainly by the major metal component. Both activity and selectivity are influenced
by conditions of temperature, pressure, and degree
of agitation.

Agglomeration of the catalyst is a problem
detrimental to the hydrogenation process, which may
be overcome by changing the pH of the medium,
changing the solvent, the solvent-substrate ratio,
or the catalyst support. Catalytic poisoning is
another problem which can affect hydrogenation.

It is difficult to predict a regeneration
technique suitable for a specific catalyst. Eventually, a catalyst reaches a point after which it
can no longer be economically regenerated.

Recycle, Recovery, and Reuse Applications:
Inorganics

Wilson (1971) describes a process in which
hydrogen is recovered from gases in which it is
concentrated at greater than 20 percent by volume.
In this process, a hydrogen-extracting aromatic
hydrocarbon is introduced countercurrently to the
system. Following hydrogenation, the aromatic
hydrocarbon is sent by countercurrent flow to a
dehydration catalyst zone. The process results in
the recovery of a hydrogen-containing gas of 90
(volume) percent.

Sulfur recovery may be effected through implementation of a process described by Rowland (1970).
Sulfur present in tail gas is converted to pure
hydrogen sulfide by hydrogenation under moderate
conditions of temperature and pressure. A cobalt
molybdate catalyst is effective in reacting water
vapor with CO to COS and CS_2 to form H_2S. Following hydrogenation and cooling, H_2S is removed and
converted to sulfur by the Stretford process, or by
recycle to the reaction furnace of a sulfur plant.

Recycle, Recovery, and Reuse Applications: Organics

Substitute natural gas may be produced from whole crude oil fractions boiling above naphtha. Components of the process include a reaction system, a quench system, gas purification and secondary hydrogenation areas, and a gas drier. Preheated oil and hydrogen gas are reacted in a fluidized bed of coke particles, resulting in the production of a gas rich in methane and ethane. The gas is quenched with a circulating stream of light aromatic to effect separation of liquid products. Hydrogen sulfide is removed from the rich gas in an absorption-stripping system and converted to elemental sulfur in a Claus plant.

In the secondary hydrogenation section, purified gas is contacted with a catalyst in a fixed bed reactor. Ethane contained in the gas reacts with hydrogen to produce methane. In cases in which there is an excess of hydrogen to ethane, carbon dioxide or hydrocarbons such as liquefied petroleum gas are added upstream of the secondary hydrogenation step, resulting in product gas containing less than 5 (mole) percent hydrogen [McMahon, 1972].

Diolefins and olefins may be selectively removed from the product gasoline obtained in light olefin manufacture, resulting in the recovery of usable hydrocarbons. Parker (1967) describes a process in which an aromatic hydrocarbon feedstock containing diolefins, mono-olefins, and sulfur contaminants is hydrogenated at a temperature of 200-500°F with a composite catalyst of lithium on palladium-alumina, to convert the diolefins to mono-olefins. The effluent is separated and an aromatic hydrocarbon is then hydrogenated at a temperature of 550-750°F with a conventional desulfurization catalyst to saturate olefins and convert sulfur compounds to hydrogen sulfide. Hydrocarbons suitable for gasoline blending and aromatic hydrocarbon suitable for solvent extraction are recovered as separate product streams.

CHEMICAL REDUCTION

Chemical reduction is a widely used industrial waste treatment process, which has the potential for recovering pollutants such as metals. This process is currently applied primarily to the control of hexavalent chromium in the plating and tanning industries.

Reduction and oxidation reactions take place concommitantly and so the overall process is referred to as an oxidation-reduction (redox) reaction. In a redox reaction, the oxidation state of at least one reactant is raised while that of another is lowered. In the following reaction:

$$2H_2CrO_4 + 3SO_2 \rightleftharpoons Cr_2(SO_4)_3 + 2H_2O \qquad (8.4)$$

hexavalent chromium (oxidation state 6+) is reduced to trivalent chromium (oxidation state 3+), while sulfur is oxidized with an increase in its oxidation state from 2+ to 3+. Sulfur compounds and base metal components such as those of iron, zinc and sodium are the more common reducing agents.

Limiting Technology

Most of the applications of chemical reduction technology to industrial waste treatment to date have been for dilute waste streams. Chemical reduction has limited applicability to slurries or sludges in their original form, because of the difficulties of achieving intimate contact between the reducing agent and the pollutant to be removed. One of the disadvantages of chemical reduction is that it may introduce new metal ions into the effluent stream. If the level of these new contaminants is high enough to exceed effluent guidelines, additional treatment will be required, adding more cost to the overall treatment.

Recycle, Recovery, and Reuse Applications

The two most common applications of chemical reduction processes are the removal of hexavalent chromium from plating and tanning industries and the removal of mercury from caustic/chlorine electrolysis cell effluents.

In plating and tanning industries, sulfur dioxide is often used for reducing hexavalent chromium to trivalent chromium, which is then precipitated as $Cr(OH)_3$ with either lime or sodium carbonate. The waste is then subjected to sedimentation, which separates the solids portions.

The reaction equations are as follows:

$$SO_2 + H_2O \rightleftharpoons H_2SO_3 \qquad (8.5)$$

$$2H_2CrO_4 + 3H_2SO_3 \rightleftharpoons Cr_2(SO_4)_3 + SH_2O \quad (8.6)$$

$$Cr_2(SO_4)_3 + 3Ca(OH)_2 \rightleftharpoons 2Cr(OH)_3 + 3CaSO_4 \quad (8.7)$$

The precipitated chromium can be potentially recovered for further use. For removal of mercury
from industrial effluents, a caustic solution of
sodium borohydride (NaBH) is mixed with the wastewater, which results in the reduction of ionic
mercury to metallic mercury. The latter precipitates out of solutions, which can then be recovered
for recycle.

Sodium borohydride is also used for reducing
lead and silver compounds from industrial effluents. These compounds are usually reduced
chemical metals, which can be precipitated, settled
and recovered for reuse.

REFERENCES

Anonymous. (1973). "Copper Industry Uses Much
    Prerap Iron." Environ. Sci. Technol.
    7(2):100-102.

Bard, A.J. (1966). Chemical Equilibrium. Harper
    and Row , New York, NY.

Butler, J.N. (1964). Ionic Equilibrium: A
    Mathematical Approach. Addison-Wesley
    Publishing Company, Inc., Reading, Mass.

Case, O.P. (1974). "Metallic Recovery from
    Wastewater Utilizing Cementation." EPA-670/2-
    74-009.

Case, O.P. (1975). "Copper Recovery from Brass
    Mill Discharge by Cementation with Scan Iron."
    EPA 670/2-75-029 #PB-241822/6WP.

Dean, J.G., F.L. Basqui, K.H. Lanouette. (1972).
    "Removing Heavy Metals from Waste Water."
    Environ. Sci. Technol. 6(6):518-22.

Federal Water Quality Administration. (1968). "A
    State-of-the-Art. Review of Metal Finishing
    Waste Treatment." Wat. Poll. Control Res.
    Ser. 12010 EIE 11/68.

Feitknecht, W. and P. Schindler. (1963).

Habarbi, F. (1970). Principles of Extractive
    Metallurgy. Vol. 2. Hydrometallurgy, Gordon
    and Breach, New York, NY.

Jancuk, W.A.D. (1976). "The Cementation Process
    for Heavy Metal Removal form Wastewater."
    Master Science Dissertation, Illinois
    Institute of Technology, Chicago, IL p. 92.

Jester, T.L., and T.H. Taylor. (1973).
    "Industrial Waste Treatment at Scovill
    Manufacturing Company." Proc. 28th Purdue
    Industrial Waste Confer. 129-137.

Keyes, H.E. (1966). "Copper Recovery Process,"
    U.S. Patent #3,288,599.

Lanouette, K.H. (1977). "Heavymetals Removal."
    Chem. Eng. Desk Book Issue. ():73-80.

Martell, A.E. and R.M. Smith. (1974a).
    "Critical Stability Constants. Vol 2: Amines."
    Plenum Press, New York, NY.

Martell, A.E. and R.M. Smith. (1974b). "Critical
    Stability Constants. Vol 3: Other Organic
    Ligands."

Martell, A.E. and R.M. Smith. (1974c). "Critical
    Stability Constants, Vol 4: Inorganic Complexes."

McMahon, J.F. (1972). "Fluidized Bed
    Hydrogenation Process for SNG."
    J. Chem. Eng. Prog. 68(12):51-54.

Parker, R.J. (1967). "Two-Stage Hydrogenation of
    an Aromatic Hydrocarbon Feedstock Containing
    Diolefins, Monolefins, and Sulfur Compounds."
    U.S. Patent #3,494,859.

Patterson, J.W. and W.A. Jancuk. (1977).
    "Cementation Treatment of Copper in
    Wastewater." Proc. 32nd Purdue Industrial
    Waste Conference 853-865.

Rowland, L. (1970). "Ninety-nine Point Nine
    Percent Sulfur Recovery Unveiled." Oilweek.
    21(32):9-12.
NG/SNG Handbook of Hydrocarbon Process. 59(4):
93-112.

Rylander, P.N. (1979). Catalytic Hydrogenation
    in Organic Synthesis. Academic Press, New
    York.

Sillen, L.G. and A.E. Martell. (1964). Stability
    Constants of Metal-Ion Complexes. Special
    Publication No. 17. The Chemical Society,
    London.

Sillen, L.G. and A.E. Martell. (1974). "Stability
    Constants of Metal-Ion Complexes." Supplement
    No. 1. Special Publication No. 25.The
    Chemical Society, London.

Stumm, W. and J.J. Morgan (1970). Aquatic Chemis-
    try: An Introduction Emphasizing Chemical Equi-
    libria in Natural Waters." Wiley-Interscience,
    New York, NY.

U.S. Environmental Protection Agency. (1974).
    "Development Document for Effluent
    Limitations. Guidelines and New Source
    Performance Standards-Iron and Steel
    Industry."

Wilson, R.F. et al. (1971). Hydrogen Recovery
    Process. U.S. Patent #3,575,690.

SUPPLEMENTAL REFERENCES

American Enka Co. (1971). "Zinc Precipitation and
    Recovery from Viscose Rayon Wastewater." U.S.
    Environmental Protection Agency, Water
    Pollution Control Research Series No. 12090
    ESG.

Arthur D. Little, Inc. (1976). "Physical, Chemical
    and Biological Treatment Techniques for
    Industrial Wastes." NTIS Publication PB-275
    287. pp. 23:1-23:33.

Cabe, V.P., B.L. Jones and R.D. Spellman. (1973).
    "Method for Simultaneous Reduction of
    Hoxanalent Chromium and Cementation of
    Copper." U.S. Patent #3,748,124.

Dean, J.G., F.L. Bosqui and K.H. Lanouette, (1972). "Removing Heavymetals from Wastewater." Environ. Sci. Tech. 6(6): 518-522.

Faigenbaum, H.N. (1977). "Removing Heavymetals in Texile Waste." Ind. Waste.

Huang, W. (1979). "Optimize Acetylene Removal." J. Hydrocarb. Proc. 59(10):131-132.

Jackson, D.V. (1972). Metal Recovery from Effluents and Sludges." Metal Finish J. 18(211); 23e, 237-8, 241-2.

Jacobi, J.S. (1966). Unit Processes In Hydrometallurgical Process. Van Nostrand Inc. Princeton, N.J.

Lanouette, K.H. and E.G. Paulson. (1976). "Treatment of Heavymetals in Wastewater." Poll. Eng.

Larsen, H.P., J.K.P. Shou and L.W. Ross. (1973). "Chemical Treatment of Metal-Bearing Mine Drainage." J. Wat. Poll. Control Ass. ():1682.

Linstedt, C.P. et al., (1971). "Trace Element Removals in Advanced Wastewater Treatment Processes." J. Wat. Poll. Control Fed. 43(7):1507-1513.

Metzner, A.V. (1977). "Removing Soluble Metals from Wastewater." Water Sewage Works. 124(4):98-101.

Monninger, F.M. (1963). "Precipitation of Copper of Iron." Min. Congr. J. 49(10):48-81.

Nadkarna, R.W., C.E. Jelden, K.C. Bowleg, H.E. Flandero and H.E. Wadsworth. (1967). "A Kinetic Study of Copper Precipitation on Iron." Trans. Mits. Soc. AIME. 239:581-585.

Peterson, R.J. (1977). Hydrogenation Catalysts. Noyes Data Corporation, Park Ridge, New Jersey.

Rickard, R.S. and M.C. Fuerstenau. (1968). "An
    Electrochemical Investigation of Copper
    Cementation by Iron." Trans. Met Soc. AIME.
    242:1487.

Scott, D.S. and H. Harlings. (1975). "Removal of
    Phosphates and Metals from Sewage Sludges.
    Environ. Sci. Tech. 9():846-855.

Steinberg, M. (1977). "Synthetic Carbonaceous
    Fuel and Feedstocks from Oxides of Carbon and
    Nuclear Power." Energy (Res. and Dev. Adm.)
    Rept. No.: CONF-770804-4 BN6-22785.

U.S. Environmental Protection Agency. (1975).
    "Development Document for Effluent Limitations
    Guidelines and New Source Performance
    Standards for the Primary Copper Smelting
    Subcategory of the Copper Segment of the Non-
    ferrous Metals.

SECTION 9

PHYSICAL DISPERSION
AND SEPARATION

INTRODUCTION

The final chapter of this report examines the recycle, recovery, and reuse applications of various physical dispersion and separation processes. The processes included in this chapter are:

1. Filtration of Liquids
2. Filtration of Gases
3. Flotation
4. Liquid-Liquid Extraction

In addition to the recycle, recovery and reuse application section for each process, a brief process description is included and, in many cases, a limiting technology statement is also given.

FILTRATION OF LIQUIDS

Filtration is a physical process, in which solids suspended in a liquid are separated from that liquid by passage through a porous medium, which separates and retains either on its surface or within itself, the solids present in the suspension. In all filtration processes, a pressure differential is induced. Depending upon the required magnitude of the pressure differential, one or more of four types of driving force may be employed: gravity, vacuum, pressure, or centrifugal. During the filter operation, a gradual pressure drop occurs due to clogging or breakthrough of suspended matter.

The filter media presently available in the commercial market can be divided into three general classes [Arthur D. Little, Inc., 1976]:

1.   A thick barrier composed of a layer of granular media such as sand, coke, coal, or porous ceramics,

2.   A thin barrier exemplified by a filter cloth or filter screen,

3.   A thick barrier composed of a disposable material such as powdered diatomaceous earth or waste ash.

The mechanics by which the particles are removed in the filters are complex. In surface filters, the predominant mechanism is usually simple mechanical straining. However, in deep-bed filters, the mechanisms can include mechanical straining coupled with gravitational settling, diffusion, interception, inertial impaction, electrostatic interactions, chemical bridging, and specific adsorption phenomena within the filter medium [Weber, 1972].

Basically, there are two types of filtration processes. Surface filters perform cake filtration, in which the solids are deposited in the form of a cake on the upstream side of a filter medium. Deep filters are used for deep-bed filtration in which solids deposit within the medium. Formation of a cake on the surface of a deep bed is undesirable [Svarovsky, 1979]. Surface filters are normally used for suspensions with more than one percent solids, whereas dilute suspensions are treated by deep-bed filters.

The filter units generally consist of a containing vessel, the filter media, structures to support or retain the media, distribution and collection devices for influent, and effluent, supplemental cleaning devices, and necessary controls for flows, water levels, or pressures. Some of the more significant alternatives in filter layout are discussed in a process design manual for suspended solids removal [U.S. EPA, 1975].

A wide variety of filtration devices are commercially available. However, they can be classified under the following categories having similar characteristics:

1.   Deep Bed Filters:

Deep bed filters were originally developed for
potable water treatment but presently are
increasingly used for industrial and municipal
wastewater treatment. The filter bed is
usually contained within a basin or tank and
is supplied by an underdrain system which
allows the filtered liquid to be drawn off
while retaining the filter medium in place.
The most common configuration of a deep-bed
filter is the downflow gravity design. The
solids concentration in the feed should be
less than about 0.1 percent by volume, in
order to keep down the number and volume of
the back wash. The most common example of
deep-bed filters is the granular media
filters, which use a bed of granular particles
(usually sand from 0.4 to 2.5 mm in size), as
the filter medium. The use of dual or multi-
media filters is becoming increasingly common.
Such filters have coarse material of low
density for the top layer and progressively
finer materials of increasing density for the
lower layers. An example of triple media is a
combination of anthracite, filter-sand, and
garnet.

Deep-bed filtration is most often operated as
a batch process. However, continuous filters
which continuously backwash a portion of the
medium are not uncommon. Some examples of such
filters are hydromation indepth filters,
radial-flow filters, and traveling backwash
filters. [U.S. EPA, 1975].

2.   Pressure Filters:

Pressure filters can treat feeds with
concentrations up to ten percent solids.
Pressure filters may be grouped into two
categories, plate and frame filter presses and
pressure vessels containing filter elements
[Svarovsky, 1979].

The conventional plate and frame press consist
of a series of plates and frames, alternately
arranged in a stack and pressed together with
hydraulic or screw-drawn rams. The plates are
covered with a filter cloth. The slurry is
pumped into the frames and the filtrate is

drained from the plates. Filter media for
plate and frame presses include various
cloths, mats, and even paper. The second
category of pressure filters includes a number
of available designs that feature a pressure
vessel containing filter elements. Some
examples in this category of filters are:
rotary-drum pressure filters, cylindrical
element filters, vertical-tank vertical-leaf
filters, horizontal tank vertical-leaf
filters, and horizontal-leaf filters
[Svarovsky, 1979]. A few of the above-
mentioned filters are used for dewatering
purposes and are discussed in a separate
report.

In addition to the filters discussed above,
there are centrifugal, fixed-bed centrifugal,
moving-bed centrifuge, and cartridge filters,
all of which are used in industrial wastewater
treatment.

Limiting Technology: Filtration of Liquids

Filtration is a well developed process and
most of the problems with this process seem to be
associated with the filter operation. The major
problem with filter operation is maintaining the
filter bed in good condition. Inadequate cleaning
results in a thin layer of compressible dirt or
flue around each grain of the medium. As the
pressure drop across the filter medium increases
during the subsequent filter run, the grains are
squeezed together, and cracks form in the surface
of the medium [Weber, 1972]. These dirty grains
gradually become larger in size and sink to the
bottom of the filter, reducing the filtration ef-
fectiveness.

Recycle, Recovery, Reuse Applications: Filtration
of Liquids

Filtration has been used for treatment of
innumerable types of industrial wastewaters.
Multimedia filtration is commonly used for removal
of the metal precipitates from wastewater after it
has been subjected to precipitation, flocculation,
and sedimentation. Filtration is also used for
dewatering of waste sludges from biological treatment systems.

FILTRATION OF GASES

Filtration is the oldest method used for removal of suspended materials in gases. It operates on the principle of passing dust-laden gases through porous filter media in which the dust is trapped. Filtration methods form one of the largest families of gas cleaning devices, and can be applied over a wide range of conditions.

All filter media collect dust by a combination of effects summarized below:

1. Particles larger than the pore size of the medium will be separated due to a sieving effect of the medium.

2. Particles may be separated by an inertial effect created by the many changes of direction the particle must undergo in passage through a tortuous filter medium.

3. Fine dust particles may be deposited on the filter medium as a result of the electrostatic charge which they often carry.

4. Sub-micron size particles, having a weight similar to that of the molecules of carrier gas, will not be separated by inertial effects. They will, instead, be subject to Brownian movement and, as a result, will be brought into close proximity to the filter medium, where they are deposited and held by electrostatic and molecular forces.

5. Particles smaller than the filter pores will be efficiently retained by a layer of dust which accumulates on and within the filter material. This layer will create a pressure drop across the filter, which must be taken into account during plant design.

Filter media can be broadly considered under three headings:

1. Gravel or Sand Aggregate Bed
2. Porous Paper and Fibrous Mats

3.   Woven and Felted Fabric Filters

Aggregate bed filters consist of uniform size
particles such as sand or gravel. These filters
are not currently in wide use for industrial gas
cleaning. Gas cleaning efficiency increases with
decreasing size of aggregate and with increasing
depth of bed. In one application, beds of fine
sand two meters deep were used to filter particles
of radioactive material from exhaust gases. These
filters were found to be over 99 percent effective
even on submicron particles.

The aggregate filter is simple in design and
can be used at elevated temperatures. When such
filters are used on gases with high moisture content, it is necessary to keep the aggregate above
dewpoint when dust-laden gases are introduced.
This prevents blockage of the bed by wet dust,
which is not dislodged by normal cleaning methods.
Fluidized beds of aggregate have generally been
found to be somewhat low in efficiency (approximately 80 percent) due to re-entrainment of dust
caused by motion of the bed.

Paper filters, due to their relatively poor
mechanical properties, are normally used at lower
temperatures where dust concentrations are less
than 5 mg-m^{-3}. These filters are not normally used
for industrial pollution control requirements.

Fibrous mat filters consist of fibers of a
natural, synthetic, or glass nature. The fibers
may be mounted in mat form between supporting
sheets of gauze or perforated metal. Filter characteristics vary according to depth of filter,
material of construction, and the density of filter
packing. Fibrous filters remove coarser material,
but when pressure losses rise due to packing density, high collection efficiencies are possible for
materials as fine as the sub-micron acid mist ori-
ginating from sulphuric acid plants.

Because of the many different kinds of filter
material available, fibrous filters are useful over
a wide range of operating variables such as temperature and corrosivity. Their application is
limited, however, to streams containing relatively
low concentrations of dust. When resistance rises
to an unacceptable level, they must be cleaned or
replaced.

Fabric filters take the form of woven fabric or felted materials manufactured from natural or man-made fibers. They are capable of treating large gas volumes with high dust concentrations on a continuous basis. The design filter rate is usually in the range of a pressure loss of 6-12 cm, depending on whether the cloth is clean or has a layer of dust deposited on it.

Woven cloth has relatively large gaps where its threads cross, which may also be large in comparison with the particles to be caught. This will result in low efficiency with clean filter bags, but as the dust builds up and blocks the holes, efficiency improves. Eventually, filtration efficiency is determined not by the filter, but by the dust layer.

Felted cloth avoids the problem of regular holes, since its fibers are laid in random fashion. The pile on its surface also increases effective filtering area, although this pile makes efficient cleaning more difficult. Synthetic fibers lack natural felting properties, and must undergo a process in which a felt is artificially induced by passing barbed needles through the fiber mat. It has been reported that needle fibers combine the high filter rate of felts with the property of being easy to clean.

The range of temperature over which fabric filters can be used is restricted. The maximum operating temperature for silicone coated glass fiber filters is 270 °C, and those for other materials commonly in use, such as teflon, nylon, wool, and cotton, are lower. New materials are being developed which are expected to extend this temperature range; stainless steel is reported to have an operating temperature which exceeds 400° C. These new materials are expensive. A further limitation on operating temperature is that the dust remains dry and flows freely. If the gas temperature falls below the dewpoint, the filtered dust will become caked on the fibers, eventually blinding the media. Removal of dust from faric filters is accomplished by flexing or collapsing the bag and blowing the accumulated dust from its surface.

Corrosive elements of gas and dust, as well as sharp edges on the dust particles, can result in rapid failure of the filter. Preferably, filter material should be chosen which has a life of several years. Another operational problem may arise when explosive mixtures are filtered, due to the electrical insulating properties of natural and synthetic fibers. If the electrostatic charge is allowed to build up, it creates an increase in voltage which may cause arcing through the gas. Explosion or fire may result. This problem can be avoided by introducing a small quantity of conducting fiber such as metal into the cloth, in order to give it anti-static properties.

Efficiency for fabric filters is always high [Parker, 1978]. Collection efficiencies in excess of 99.5 percent are common. Fabric filters are commonly used for control of dust concentrations in the range of 10^3 µg/m^3 (urban atmospheric dust) to 10^3 µg/m^3 (pneumatic conveying). They provide effective removal of particles ranging in size from submicrometer fumes to 200 m powders [Billings, 1977].

Limiting Technology

Filter media used for removal of suspended materials in gases can be broadly grouped into three categories: gravel or sand aggregate beds; porous paper and fibrous mats; and woven and felted fabric filters.

The aggregate filter is simple in design and can be used on dust-laden gases with high moisture content. It is necessary to keep the aggregate above dewpoint to prevent blockage of the bed by wet dust, which is not dislodged by normal cleaning methods. Fluidized aggregate beds may have low efficiency due to re-entrainment of dust caused by motion of the bed.

Paper filters have poor mechanical properties, and are not normally used for industrial pollution control.

Fibrous filters are useful over a wide range of operating conditions such as temperature and corrosivity, due to the many types of filter material available. Their application is limited to streams containing relatively low dust concentrations.

Fabric filters are capable of treating large gas volumes with high dust concentration on a continuous basis. With woven cloth filters having large holes, filtration efficiency is determined ultimately by the dust layer which collects on the filter, and not by the filter itself. The use of felted cloth eliminates the problem of regular holes, since its fibers are laid in random fashion.

The temperature range for the use of fabric filters is restricted by the type of filter material. New materials are being tested which are expected to extend the range of operating temperatures to 400°C and above. The gas temperature must stay above dewpoint to eliminate the problem of dust caking on the filters.

Corrosive elements of gas and dust, and sharp dust particle edges, can cause rapid filter failure. An additional operational problem is the build-up of electrostatic charge on the filter. This must be prevented in order to prevent explosion of gas or an outbreak of fire. Introduction of a small quantity of conducted fiber into the filter cloth will give it antistatic properties. Collection efficiencies in excess of 99.5 percent are normal for fabric filters.

Recycle, Recovery, and Reuse Applications: Organics

Due to the shortage of oil, the recycling of asphaltic material has become an issue of interest. A problem which occurs in trying to convert conventional asphalt plants to plants which have recycle capabilities is the formation of fine particulate smoke which occurs when crushed asphalt pavement is subjected to temperatures necessary for recycling. At these temperatures, the asphalt begins to crack and release hydrocarbon vapor which condenses into submicron droplets. These oil droplets are poorly collected by baghouse filters and venturi scrubbers, devices currently in use at asphalt plants.

The electrofluidized bed (EFB), designed for high efficiency collection of submicron particles, was tested on emission from an asphalt recycling plant. The hydrocarbon pollutant is collected on sand, which is then removed from the bed in its fluidized state and added to the asphalt product. Efficiencies of collection in excess of 98 percent were reported for submicron particles in beds

having unfluidized depths of 8-12 cm, and using
sand having a mean diameter of 2 mm. Using this
procedure, it was possible to combine collection of
submicron particles at high efficiency with the
recycling and reuse of the collected hydrocarbons
and the filter sand [Zieve, et al. 1978].

Recycle, Recovery, and Reuse Applications:
Inorganics

    In certain manufacturing operations involving
expensive water insoluble dusts or powders in which
excess material is normally exhausted as part of
the operation, it has been found to be economical
to collect exhaust material in a filter, from which
the retained materials can be recovered. Such a
method is used for collecting phosphor particles
produced in forming color cathode ray tubes by the
dry phosphor technique.

    Excess phosphor particles are collected from
the exhaust atmosphere by passing exhaust air
through several filter media. The first filter is
normally coarse woven, and is cleaned by vacuuming,
shaking, or beating. The remaining superfines are
then collected by filters made of materials such as
cellulose, glass, or plastic in the form of fibrous
compactions. These filters are not suited to mechanical removal of phosphor without destruction.
In some cases, the phosphor particles have been
dissolved for reclamation by organic solvents or
similar chemicals. This is an expensive process,
however, and caustic solvents may be detrimental to
the phosphor [Warner, 1971].

    Warner (1971) developed a decomposable filter
for collecting water in soluble particles of gas-
borne materials. The filter medium is made up of
water soluble organics which are stable in an environment having ambient relative humidity less
than 90 percent and an ambient temperature under
100 $^\circ$C. Particulate material is reclaimed by dissolving the filter in water. The particulate matter is allowed to settle, and is recovered by
decanting the supernatant. Alternatively, the
mixture of particulate and dissolved filter medium
is passed through a recirculating filter having a
core of perforated discs upon which the particu-
lates are collected. Heated air is then forced
through the filter in a reverse manner to dislodge
and dry the collected particulate.

FLOTATION

Flotation is a unit operation used to separate solid or liquid particles from a liquid phase. Separation is brought about by introducing fine air bubbles into the liquid phase. The bubbles attach to the particulate matter and the buoyant force of the combined particle and gas bubbles is large enough to cause the particle to rise to the surface [Metcalf and Eddy, 1980]. Three methods of introducing gas bubbles have been shown to create bubbles sufficiently fine for flotation of suspended solids in municipal and industrial wastewaters.

1.   Injection of air while the liquid is under pressure, followed by release of the pressure (dissolved air flotation).

2.   Aeration at atmospheric pressure through revolving impeller or porous media (induced air flotation).

3.   Application of a vacuum to the wastewater, which is saturated with air at atmospheric pressure.

In biological treatment systems, biological flotation occurs when the gases formed by natural biological activity are attached to the suspended solids and rise upwards through the liquid. For any of these systems, the degree of removal can be enhanced through the use of various chemical additives [Metcalf and Eddy, 1980].

Dissolved Air Flotation

In dissolved air flotation (DAF) systems, bubbles of size 100μ or less are obtained by dissolving air in water at an elevated pressure and then reducing the pressure of the air-water mixture. The water is pressurized in the range of 40-80 psi and a stream of air is injected into the pressurized water and retained in a tank under pressure for several minutes to allow time for the air to dissolve. At $60°F$, about one cubic foot of air is used for each 100 gallons of air-charged water [U.S. EPA, 1975]. The air-charged water then passes through a pressure reducing valve into the flotation treatment tank. The reduction in pressure causes the solubility of the air to decrease and excess air comes out of solution in the form of

minute bubbles (average size, 18µ). These bubbles
attach to the solid particles and increase their
bouyancy, causing them to float to the surface. In
some cases a portion of the effluent (15-120 per
cent) is recycled, pressurized and semi-saturated
with air [Metcalf and Eddy, 1980]. A coagulant aid
may be used with a DAF unit in order to 1) increase
the allowable solids loading; 2) increase the percentage of floated solids; and 3) improve the clarity of the subnatant.

Induced Air Flotation

    In induced air-flotation systems, coarse air
bubbles are formed in the flotation tank by a
revolving impeller or through diffusers. These
systems are not common in industrial wastewater
treatment operations, due to the low efficiency of
the coarse bubbles formed.

Vacuum Flotation

    In vacuum flotation systems, the wastewater is
initially saturated with air and a partial vacuum
is then applied which causes the dissolved air to
come out of solution as minute bubbles. These
bubbles attach to the solid particles in wastewater
and rise to the surface to form a scum blanket,
which is easily removed by a skimming mechanism.

Recycle, Recovery and Reuse Applications

    Flotation is a process initially developed for
ore concentration and used principally by the mineral industry. In wastewater treatment it is mostly used to remove suspended matter and concentrate
biological sludges. There have been other applications of flotation in the waste treatment area,
however.

    In a variation of the process called precipi-
tate flotation, metal precipitates can be collected
on the top of the flotation tank and recovered as a
froth concentration [Arthur D. Little, Inc., 1976].
In this application, the metal to be removed from
solution is first precipitated. Another possible
application of precipitate flotation is the flotation of complexed cyanide (for example, ferrocyanide
precipitate), as a means of removing cyanide from
solutions [Grieves and Bhattacharya, 1969].

In another variation of the flotation process, known as ion-flotation, a surfactant ion of opposite charge to the inorganic ion removed from solution is added in stoichiometric amounts. The surfactant, which must exist in solution as simple ions, reacts with the inorganic ion to form an insoluble "soap" which is raised to the surface with a bubbling action. Rubin et al. (1966) reported removal of dissolved copper with sodium lauryl sulfate using the ion flotation process.

The insoluble metal precipitates formed during ion and precipitate flotation processes can be further subjected to treatment for recovery of metals.

The flotation process is used for recovery of copper from the slag generated in smelters [Matheson, et al. 1976]. Flotation is also used for removal of oil from industrial wastewaters such as those from refineries.

LIQUID-LIQUID EXTRACTION

Liquid-liquid extraction has become an important separation technique used by various industries either to remove small amounts of an impurity from a product stream or to separate products. This technique is considered to be a viable waste treatment process for selected waste streams, where recovery of material is possible. The major applications of this process in wastewater treatment engineering are 1) recovery of phenol and related compounds from wastewaters and 2) removal of water soluble solvents such as alcohol from wastes containing mixed chlorinated hydrocarbon solvents.

Liquid-liquid extraction, hereinafter referred to as solvent extraction, involves separating the components of a liquid mixture by the addition of another liquid referred to as the solvent which is immiscible (or only partially miscible) with the initial phase. The solvent is chosen so that one or more of the components of the original solution, called the solute, will transfer preferentially into the solvent phase, leaving the others behind in the so-called "raffinate." The product of the desired solute in the solvent is called the extract. At equilibrium, the ratio of the concentrations of solute in the extract, y, and raffinate, x, phases is called the distribution coefficient,

D. The coefficient is used as an indicator of the ability of a solvent to extract a particular solute.

The proportion of solute recovery in a single equilibration depends on both the distribution coefficients for that solute and the relative amount of solvent used. To obtain a high recovery without using an excessive amount of solvent (with the corresponding production of a very dilute extract), most processes employ multistage countercurrent contacting [Hanson, 1979].

The purity of the initial extract is often enhanced by scrubbing with a suitable immiscible phase so that the bulk of the desired solute in the solvent phase is retained while the impurities are washed out. The scrub feed is usually based on the same solvent as the original feed and the scrub raffinate is then combined with the main feed to permit recovery of any of the desired solute which may have transferred.

The solute may be removed from the extracted solvent by a second solvent extraction step, distillation, or an alternative process. Depending upon cost consideration, solvent recovery from the treated stream may be advantageous. This may be accomplished by stripping, distillation, adsorption, or other suitable processes.

Various devices have been developed to establish phase equilibrium rapidly and thereby accomplish solvent extraction based on a combination of capital and operating costs. Several discussions [Oberg and Jones, 1963; Bailes, et al. 1976; Reissinger and Schroeter, 1978a, 1978b] on the various extraction devices in use are available in the literature.

The extraction devices have two things in common; first, they generate a large amount of interfacial area between the two liquid phases, mostly by dispersing one phase in the other and, secondly, they impart mechanical energy into the system by means of agitation to maintain a degree of turbulence in one or both phases [Ecker, 1976]. A thorough discussion of the different types of extractions is given by Reissinger and Schroeter (1978a, 1978b).

There is no general guide to contactor selection. Most of the contactors available in the market may be used for simple processes and a final choice will depend on the results of an economic evaluation of available options as well as the consideration of local factors such as space availability [Bailes, et al. 1976]. Reissinger and Schroeter (1978a, 1978b), have attempted to establish a selection procedure based on the use of a diagram into which the most important parameters have been incorporated. The suggested scheme may only be used as a rule of thumb. The final design of specific solvent extraction unit of a new application should be preceded by detailed laboratory tests as well as technical and economic comparisons with competing processes.

Limiting Technology

Liquid-liquid (solvent) extraction is a well established process which has relatively few insurmountable technical problems. The problems associated with the application of solvent extraction to recovery of by-products from waste streams are basically related to the difficulty in selecting solvents and a contactor, which are, in combination, capable of producing the desired results.

Recycle, Recovery and Reuse Applications

Liquid-liquid extraction is being used in both commercial processing and waste applications. The following are the major waste treatment applications:

1.   Removal and recovery of phenol and related compounds from petroleum refinery waste [Anonymous, 1973; Coke-oven liquors, Carbone, 1967; Aver et al. 1969) and phenol resin plant effluents [Wurm, 1968].

2.   Removal and recovery of water-soluble solvents such as alcohol from wastes containing chlorinated hydrocarbon solvents.

3.   Extraction of thiazole-based chemicals [Anonymous, 1970], acetic acid [Himmelstein, 1974] and Salicylic acid [Anonymous, 1970].

REFERENCES

Anonymous. (1970). "Petrochemical Effluents
    Treatment Practices." Fed Water Poll. Control
    Adm., U.S. Department of the Interior. Report
    No. 120120. p. 58.

Anonymous. (1973). "Manual on Disposal of
    Refinery Wastes-Volume on Liquid Wastes." In:
    Stripping, Extraction, Adsorption and Ion
    Exchange. American Petroleum Institute,
    Washington, D.C.

Arthur D. Little, Inc. (1976). "Physical,
    Chemical and Biological Treatment Techniques
    for Industrial Wastes." NTIS Publication PB-
    275 287. pp. 22:1-22:27.

Bailes, P.S., C. Hanson and M.A. Hughes. (1976).
    "Liquid-Liquid Extraction: The Process, The
    Equipment." Chem. Eng. ():86-100.

Billings, Charles E., (1977). "Fabric Filter
    Installations for Flue Gas Fly Ash Control."
    Status Report Powder Technol. 2-18 (1):70-110.

Carbone, W.E., R.N. Hall, H.R. Kaiser and G.C.
    Bazell. (1958). "Commercial Dephenolization
    of Ammoniacal Liquors with Centrifugal
    Extractors."

Eckert, J.S. (1976). "Extraction Variables
    Defined" Hydro. Proc. ():117-124.

Grieves, R.B. and D. Bhattachanya. (1969).
    "Precipitate Flotation of Complexed Cyanide."
    Proc. 24th Ind. Waste Conf., Purdue
    University, p. 8.

Hanson, C. (1979). "Solvent Extraction-An
    Economically Competitive Process."
    Chem. Eng. (New York). 86(10):83-87.

Himmelstein, K.J. (1974). "Removal of Acetic Acid
    from Wastewaters." Proc. of the 29th Ind.
    Waste Conf., Purdue University. pp. 677-685.

Lauer, F.C., E.J. Littlewood and J.J. Buttler. (1969). "New Solvent Extraction Process for Recovery of Phenols from Coke Plant Aqueous Waste." Paper presented at Eastern States Blast Furnace and Coke Oven Association Meeting, Pittsburg, PA.

Matheson, K.H., Jr., R.R. Beck, R.J. Heaney and C.K. Lewis. (1976). "Emission Control Effort at Kennecott's Utah Smelter." AIChE symp. Ser. 72(156):312-320.

Metcalf and Eddy, Inc. (1980). Wastewater Engineering Treatment Disposal, and Reuse. 2nd ed. McGraw Hill Book Co., New York, NY.

Oberg, A.G., and S.C. Jones. (1963). "Liquid-Liquid Extraction." Chem. Eng. 70(15):119-134.

Reissinger, K.H. and J. Schroeter. (1978a). "Modern Liquid-Liquid Extractions: Review and Selection Criteria." Inst. Chem. Eng. Symp. Ser. No. 54.

Reissinger, K.H. and J. Schroeter. (1978b). "Selection Criteria for Liquid-Liquid Extractions." Chem. Eng. (New York). 85(25):109-118.

Rubin, A.J., D.J. Johnson, and J.C. Lamb. (1966). "Comparison of Variables in Ion and Precipitate Flotation." Ind. Eng. Chem. 5(4):368-375.

U.S. Environmental Protection Agency. (1975). "Process Design Manual for Suspended Solids Removal." US EPA Technology Transfer Report No. EPA 625/1-75-003a.

Warner, J.G. (1974). Decomposable Filter Means and Methods of Utilization, U.S. Pat. 3,616,603 Nov.

Weber, W.J., Jr. (1972). Physicochemical Processes for Water Quality Control. Wiley Interscience, New York, N.Y. pp. 139-142.

Wurm, H.J. (1968). "The Treatment of Phenolic Wastes." Proc. of the 23rd Ind. Waste Conf., Purdue University. pp. 1054-1073.

Zieve, B., Karim, Z., and Melcher, R.J. (1978).
"Electrofluidized Bed in Pavement Recycling
Process," J. Env. Science & Technology, Jan.
12(1):4.

SUPPLEMENTAL REFERENCES

Abbott, J.H., Drehmel, D.C., (1976)., "Control of
Fine Particulate Emissions." Chem. Eng. Prog.
72(12):47-51.

Albert, J.T. (1977). "Waste Activated Sludge
Thickening by Countercurrent Flotation."
Proc. 32nd Ind. Waste Conf., Purdue
University, pp 1.

Bakhai, Narendra N., et al. (1975). "Treatment of
Tar Sands Tailings With Fly Ash; Env. Sci.
Technol 9(4):363-364, April.

Chem. Eng. Prog. (1966). 62(9):49-104. Contains
9 articles on Solvent Extraction.

Cleasby, J.L. (1976). "Filtration and Separation:
Filtration with Granulan Beds." Chem. Engr.
():663-667, 682.

Dave, G., Blanck, M. and Gustaffson, K., (1979).
"Biological Effects of Solvent Extraction
Chemicals on Aquatic Organisms." J. Chem.
Technol. Biotechnol. 29(4):249-257.

Davis, J.C. (1971). "New Technology Revitalizes
Waste-Lube-Oil Refining." Chem. Eng.
():62-63.

Earhart, J.P. and King, C.J. (1976). "Recovery of
Essential Oil and Suspended Solid Matter from
Lemon Processing Wastewater by Volatile-
Solvent Extraction and Emulsion Flotation."
J. Food Sci. 41 ():1247-1248.

Earhart, J.P., Won, K.W., Wong, H.Y., Prausnitz,
J.M. and King, L.J. (1977). "Waste Recovery:
Recovery of Organic Pollutants via Solvent
Extraction." Chem. Eng. Prog. ():67-73.

Fitch, B. (1977). "When to Use Separation
Techniques Other Than Filtration." AIChE
Symp. Ser. 73(171):104-108.

Grutsch, J.F. and Mallatt, R.T. (1977).
    "Filtration and Separation: Optimizing
    Granular Media Filtration." Chem. Eng.Prog.
    ():57-66.

Gulman, C.S. and Baumann, E.R. (1977). "What,
    When and Why of Deep Bed Filtration." AIChE
    Symp. Ser. 73(171):76-82.

Gusmer, J.H. (1977). "Asbestos Containing Filter
    Materials." AIChE Symp. Ser. 73(171):33-37.

Hanson, C. (1971). "Recent Advances in Liquid-
    Liquid Extraction." Pergamon Press, New York,
    NY.

Holeson, M.J. (1976). "Review of Baghouse Systems
    for Boiler Plants," J. Air Pollution Control
    Assoc. 26(1):22-26.

Hutto, Jr., F.B. (1977). "What the Filterman
    Ought to Know about Filteraid Filtration."
    AIChE Symp. Ser. 73(171):50-54.

Iammartino, Nicholas R., (1972). "Technology
    Gears Up to Control Fine Particles. J. Chem.
    Eng. 21, 79 (18):50.

Janoso, Richard P., Meyler, J.A., (1976).
    "Baghouse Operating Experience With Coal
    Firing." Power Eng. 80(10):62-64.

Katz, W.J. and Geinopolos, A. (1967). "Sludge
    Thickening by Dissolved Air Flotation."
    J. Water Poll. Control Fed. 39(6):946.

Keel, Kevin K., (1977). "Equipment and Techniques
    for Removing Particulates From High
    Temperature Gas Stream," Plant Eng.
    31(10):111-115.

Kempling, J.C. and Eng, J. (1977). "Performance
    of Dual-Media Fitlers-2." Chem. Eng. Prog.
    ()87-91.

Kiezyk, R.R. and Mackay, D. (1971). "Wastewater
    Treatment by Solvent Extraction." Can. J.
    Chem. Eng. 49(6):742-752.

Kiezyk, R.R. and Mackay, D. (1973). "Screening and Selection of Solvents for Extraction of Phenol from Water." Can. J. Chem. Eng. 51():741-745.

Kinsman, S. (1977). "Instrumentation for Filtration Tests." AIChE Symp. Ser. 73(171):13-17.

Kohn, P.M. (1976). "New Extraction Process Wins Acetic Acid from Waste Streams." Chem. Eng. ():58,60.

Komline, T.R. (1978). "Dissolved Air Flotation Tackles Sludge Thickening." Wat. Waste Eng. 15(2):64.

LeClerc, G. (1971). "Surface Treatment Effluents. Problems and Solutions." Text in French. Galvano (Paris), 40 (407):39-46, Jan. 1971.

Lloyd, P.J. and Ward, A.S. (1977). "Filtration Applications of Particle Characterization." AIChE Symp. Ser. 73(171):6-12.

Luthy, R.G., Selleck, R.E. and Galloway, T.R. (1978). "Removal of Emulsified Oil with Organic Coagulants and Dissolved Air Flotation." J. Wat. Poll. Control Fed. 50(2)331-346.

Macey, L.J., et al. (1980). "Low-Temperature Approach to Airborne Pollution Control," Chem. Eng. 355, pp. 216-218.

Mattila, T.K. and Lehta, T.K. (1977). Nitrate Removal from Waste Solutions by Solvent Extraction." Ind. Eng. Chem. Process Res. Dev. 16(4):469-472.

Mayhue, L.F. (1972). "Solvent Extraction Status Report." USEPA Report No. PB-221 458/3.

Mohler, Jr. E.F., and Clere, L.T. (1977). "Filtration and Separation: Removing Colloidal Solids Via Up-flow Filtration." Chem. Eng. Prog. ():74-82.

"New Air Filtration Process Saves Roofing Plant in
    Pollution Crises," (1972) <u>2nd Warkes</u>, 18
    (5):pp 54-55, Siftlock.

Nickolaus, N. (1977). "The What, When and Why of
    Cartridges." AIChE Symp. Ser. 73(171):38-49.

Olson, R.L., Ames, R.K., Peters, H.H., Gustan,
    E.A., and Bannon, G.W. (1975). "Sludge
    Dewatering with Solvent Extraction." Paper
    Presented at the Natl. Conf. on Management and
    Disposal of Residues from the Treatment of
    Industrial Wastewater, Washington, D.C.

Parker, A.J. (ed). (1978). <u>Industrial Air-
    Pollution Handbook</u>. McGraw-Hill, U.K.

Ramirez, E.R. (1979). "Comparative Physiochemical
    Study of Industrial Wastewater Treatment by
    Electrolytic, Dispersed Air, and Dissolved Air
    Flotation Technologies." Proc. 34th Ind.
    Waste Conf., Purdue University, pp. 699.

Reves, Sidney M., (1979). "New Scrubber Process
    is Sludge-Free." <u>Coal Min Process</u>
    16(16):50-54.

Ricker, N.L., Michaels, J.N., and King, C.J.
    (1979). "Solvent Properties of Organic Bases
    for Extraction of Acetic Acid from Water."
    <u>J. Separ. Proc. Technol</u>. 1(1):36-41.

Ricker, N.L., Pittman, E.F., and King, C.J.
    (1980). "Solvent Extraction with Amines for
    Recovery of Acetic Acid from Dilute Aqueous
    Industrial Streams." J. Separ. Proc. Technol.
    1(2):23-30.

Schmid, B.K., Jackson, D.M. (1979). "Liquid and
    Solid Fuels by Recycle SRC." <u>J. Coal Process
    Technol.</u> V5, published by MICHE, p. 146-15.

Scheibel, E.G. (1954). "Calculation of Liquid-
    Liquid Extraction Processes." <u>Ind. Eng. Chem.</u>
    46().

Scheibel, E.G. (1956). "Performance of an
    Internally Baffled Multi-stage Extraction
    Column." <u>AIChE J</u>. 2():74-78.

Shelozukhov, D.A., et al. "Purification of Mercury
    Containing Gases in Tabular Furnaces From Dust
    in Dry Electro Filters. Text in Russian.
    Tsvetn. Mebal., 43(1):35-39.

Stern, Sidney C., (1978). "Mechanisms and
    Materials For Fabric Dust Filtration,"
    Proc. on The Int. Fabric Alternatives Forum,
    3rd, Phoenix, Ariz., published by Am. Air.
    Filter Co., Inc., Louisville, KY, p.4.1-4.19.

Suttle, H.K. (1969). "Filtration." Grampian
    Press, London.

Treybal, R.E. (1963). Liquid Extraction. 2nd
    Edition. McGraw Hill Book Company, New York,
    NY.

Upmalis, Alberto, (1972). "Experiment With a Wet
    Filter For Heat Recoery From Flue Gases."
    Text in German, Brennstoff Walrine Kraft,
    24(5) pp 208-211.

Van Turnbout, J.C., Van Bochove, G.J., Van
    Veldhimzen, (1976). "Electret Fibres For High
    Efficiency Filtration of Polluted Gases,"
    Staub Reinhalt Luft 36(1):36-39.

Vrablik, E.R. (1959). "Fundamental Principles of
    Dissolved Air Flotation of Industrial Wastes."
    Proc. of 14th Ind. Waste Conf., Purdue
    University. p. 743.

Wardell, J.M. and King, C.J. (1978). "Solvent
    Equilibria for Extraction of Carboxylic Acids
    from Water." J. Chem. Eng. Data. 23(2):144-
    148.

Witt, P.A., Jr., and Forbes, M.C. (1971).
    "Valuable By-Product Recovery by Solvent
    Extraction." AIChE Symp. Ser.-Water.
    68(124):108-114.

Woods, D.R. (1973). "Treatment of Oily Wastes
    from a Steel Mill." J. Wat. Poll. Control Fed.
    45(10):2136-2145.

World Filtration Congress, 1st, Papers, (1974).
    Published by Halsted Press, Div. of John Wiley
    & Sons, New York, NY.

Zeitoun, M.R., Davidson, R.R., and Wood, D.W.
    (1966). "Renovation of Sewage Plant Effluents
    by Solvent Extraction." US EPA Report
    No. PB-230 080.